Understanding Maps

Understanding Maps

Second Edition

J.S. Keates

Longman

Addison Wesley Longman Limited
Edinburgh Gate, Harlow
Essex CM20 2JE, England
and Associated Companies throughout the world

© Longman Group Limited 1982
© This edition Addison Wesley Longman Limited 1996

First edition 1982
Second edition 1996

British Library Cataloguing in Publication Data
A catalogue entry for this title is available from the British Library.

ISBN 0-582-23927-3

Library of Congress Cataloging-in-Publication data
A catalog entry for this title is available from the Library of Congress.

Set by 7 in Palatino 10/11 pt
Produced through Longman Malaysia, LWP

Contents

CONTENTS

CONTENTS

List of figures

Preface to the first edition

In a book dealing with maps and visual perception it would be desirable to include a number of reproductions of actual maps. Apart from the technical difficulties of reproducing them to their individual colour specifications, it would be prohibitively expensive. Consequently the reader is requested to refer where necessary to some of the types of map which are commonly available: for example, a city street plan, a road map, a medium-scale topographic map (such as 1 :50 000), and a general atlas. If any specialised maps, such as a geological map and a coastal nautical chart, are also available, then so much the better.

These essays cannot examine every possible aspect of cartography and cognition. The central theme, therefore, is to identify the principal factors in visual perception which affect both making and using maps, for although they operate with different objectives, they are both essentially visual. The first part of the book is an attempt to apply the knowledge derived from physiology and psychology to those aspects of visual perception which are most relevant to maps, and at the same time to focus attention on the specific visual characteristics of maps.

The second part deals with the signs and symbols used on maps, considering them from the point of view of sign systems in general. This is a matter strangely neglected in cartographic studies. Yet it is particularly important to initiate such a discussion, because the functions and limitations of the particular sign system are critical in understanding how maps work.

The third part faces the problem of the relationship between the map maker and the map user. Although this topic is widely discussed under the heading of 'cartographic communication', generally speaking the supporting evidence for the general theories advanced is lacking. A critical review leads to the conclusion that at present any such general theory is a long way from realization. Even so, it is hoped that these essays will illuminate some of the factors which must be taken into consideration in pursuing the objective of understanding maps.

J.S. Keates   Glasgow   January 1981

Preface to the second edition

Whereas in the 1970s the impression given by many publications was the desirability of replacing the 'subjective' cartographer by operations based on properly verified scientific principles, by the late 1980s there was an emerging recognition of the need to learn from and indeed incorporate cartographic expertise into the development of artificially intelligent and expert systems applied to map production. This itself was largely consequent on the rapid development of high-quality digital cartographic and geographic information systems.

Although such developments are seen primarily as applied technology, they continue to raise fundamental questions about the nature of the map, the relationship between maps and other graphic images, and the respective roles of cartographer and map user. The first edition considered problems of visual perception in relation to cartography and map use; the nature of map symbols; and the theories of cartographic communication which were proposed as models. This second edition extends this to other topics which either have been neglected entirely – such as cartographic skill and the connections between cartography and art – or which could benefit from being examined from different points of view, such as the rhetorical nature of the map and the relationship between cartographic theory and practice. For example, cartographic generalization, which as a basic practice is covered by many texts, is looked at under three headings: from the point of view of the understanding of the map user; in relation to the practices of drawing and painting; and as an example of cartographic skill.

Inevitably, the question of illustration is difficult. Generally speaking, the policy adopted here is not to reproduce figures that are available in existing cartographic publications. So far as maps and atlases are concerned, the problems are even greater, for including small sections of atlas pages or sheet maps does not do justice to the original materials or the problems of map use, even if facsimile reproduction could be guaranteed. In addition, the object of this book is not to invite the reader to scrutinize particular works, but to apply the ideas presented here to the maps and atlases which the reader encounters.

Once again, it should be emphasized that this is neither a text book nor a survey of cartographic literature. The ideas presented here are offered simply as a basis for discussion.

J.S. Keates   Glen Esk March 1995

Acknowledgements

Most of the figures were prepared by Każia Kram and Mike Shand, with photographic processing by Leslie Hill, and their assistance is gratefully acknowledged.

We are grateful to the following for permission to reproduce copyright material:

Fig. 30, representation of a thunderbird in West Coast American Art from *Kwakiutl Ethnography* by Hans Boas published by University of Chicago Press © 1966 by The University of Chicago. All rights reserved. Reprinted with permission.

Whilst every effort has been made to trace the owners of copyright material, in a few cases this has proved impossible and we take this opportunity to offer our apologies to any copyright holders whose rights we may have unwittingly infringed.

PART ONE

The map as visual information

CHAPTER 1

Using maps

To most people, using a map appears to be a straightforward and quite natural task. Finding a place; selecting a route; discovering the name of a prominent hill; all these are common and indeed, quite elementary. It is also clear that some types of map use are more complex; for example, trying to memorize a route, or navigate a ship in a narrow channel. Yet even these would seem to have some basic elements in common. People use maps to obtain information, to find something out, and they are aware, consciously or subconsciously, that for some purposes the map is far more effective than any other source.

At the same time, most map users will admit that on occasions they have found maps to be unsatisfactory. They discover, after a fruitless search, that the place they are seeking is not shown, that the information on the map is 'out of date', or that the map is 'difficult to read'. Few go beyond this point and attempt to find out what has gone wrong, and in virtually all cases the map (or the map maker) will be blamed, rather than their own incompetence, wrong judgement, or lack of skill. Indeed, many users would maintain that using a map should require no more than normal vision and average intelligence – which is interesting when compared with the number of years of formal education devoted to the other primary means of communication in our society, language and mathematics.

It is not difficult to show that the apparently simple act of 'looking at a map' is highly complex, and involves a whole sequence of processes, not all of which are perfectly understood. This can best be illustrated by taking a simple map-using task, and attempting to account for what actually takes place in the process of obtaining the desired information.

A hypothetical example

Let us suppose that our map user wants to find the location of a small river, which he intends to visit, in order to go fishing. His first problem is to decide what map to look at, or to find what he believes to be a suitable one. In many cases the result will be largely accidental: he looks at the maps he has, whether they are likely to contain the desired information or not. Indeed, failure to obtain the information is as likely to result from this as from any other cause. However, he may pick up a road atlas, or a general atlas he happens to have in the house. It may well turn out that the river he is seeking does not appear at all on the general atlas map, nor can he find the name in the road atlas. His reaction may be that 'the maps are no good' and he may discontinue the search. On the other hand, he may persist, and obtain a larger-scale topographic map of the area in which he knows or believes the river to be.

As there is no name index, he has to search for the river or the river name. How long this takes will depend on many factors, including the relative size and importance of the river, and his existing geographical knowledge. Nevertheless, he is looking for something, and either he knows beforehand, or learns as he goes along, that what he is seeking is a blue line and a 'blue' name.

At this point we can consider what has happened so far. In the first place there is not simply one map of the area concerned, and those that the user has at his disposal are a few out of possibly a large number that have been made, or could have been made. The atlas map does not have the information because it is too small in scale, and the smaller the scale of the map the greater is the chance that most 'small' features will have been omitted. The road map does not name the river, because it concentrates on roads and motoring information, and therefore the names of minor rivers are unimportant. Maps, therefore, have to be considered in terms of scale and content, and their use depends on judgement. All maps are selective. The first problem in any map-using task is to find out whether a suitable map exists; very often the second problem is the difficulty of obtaining it.

In trying to discover the location of the river, the map user is endeavouring to extend his knowledge of the real world in a matter which at this time is important to him. This is an aspect of cognition, which Neisser (1976: 1) has defined as 'the activity of knowing: the acquisition, organization and use of knowledge'. The blue line he is looking for on the map is a symbol. Symbols are a subset of signs, and the use of signs is fundamental in human communication and expression. Whereas cognition is concerned mainly with the psychological aspects of how we learn, think and remember, the same problems are important in philosophy, and in particular epistemology, which deals with the origins and methods of knowledge. How we are able to think and learn about the external world is of course a central problem in philosophy.

The map user can learn about this particular river in two ways. He can visit it, walk along its banks, and experience it directly; or he can make use of a description of it, in this case through a map, and thereby learn about it indirectly. For all human beings, knowledge is in part acquired through direct experience, that is, by perception of the external world through the senses; and in part by indirect experience, making use of some form of communication.

One advantage of using a map is of course that it is a store of information, and therefore it can be consulted without actually visiting the river. Consequently it enables the user to anticipate a future requirement. On the other hand, the amount of information which can be gained from the graphic image of the map is not at all the same as that which would result from direct experience.

The blue line on the map represents the concept 'river', the general term we use to describe all bodies of water which flow along particular courses. The process of placing all external objects, features or events into classes or categories is fundamental to language, thought and learning. Therefore, this particular river in which the map user is interested is not described individually, but simply placed within the category 'river'. It is evident that a basic principle in the map-using process is that both the map maker and the map user are able to agree on the meaning of the term 'river'.

Although in this respect the map information is limited, in another respect it is quite specific. It shows the location of the river on the Earth's surface; its position, course and extent; and its relationship in space to other features. Because the map is a miniature representation, the map user can perceive the whole of the river and its physical relations, which in most cases he would find difficult, if not impossible, to do by direct experience. This part of the map information is not conceptual, but factual. Because the map is a two-dimensional graphic image, it provides this locational information systematically. No other river can be in this particular place, and the map states this without any ambiguity. Therefore the map, despite its inherent limitations, both in the range of information which can be included without confusion, and its very limited ability to show quantities graphically, is the most effective means of recording and communicating locational information of this type. In this respect it is complementary to, and not a substitute for, the other principal means of communication, which are verbal and sequential.

In the process of searching for the river, the map user is looking at the map, and looking for the blue line with the particular name. The next series of questions is concerned with what actually takes place during this part of the map-using operation. Even if he has little idea of where the river may be, of its size and length in relation to map scale, he is unlikely to search at random. He will make use of the visual clues provided by the map symbols (the blue lines of the rivers being quite distinct from the brown lines of the contours); by the geographical evidence of likely or unlikely locations (he will not dwell on large bodies of water or large urban areas); and by any other clues which may be remembered from what other people have told him. The ability to respond to the visual structure of the map, the ability to make inferences from geographical evidence, and the amount of geographical knowledge which can be recalled will all vary between different users. The process of acquiring knowledge is cumulative, and every person builds up and makes use of existing knowledge and experience. Finding a place on a map, therefore, is affected by what the user knows in advance. This applies, of course, not only to obtaining information from maps but to all information-gathering processes.

Obtaining information from a map by looking at it involves visual perception. This complex process is initiated when light from the object perceived (in this case part of the map) forms a pattern on the retina of the eye. This initial stimulus (which is a physical flow of energy) results eventually in a psychological response. The receptor cells in the retina send signals through the optic nerves which eventually reach the brain. Somehow the brain converts the physical stimuli into meaning, so that the signals which indicate a blue line on the map are converted into the response 'river'. Signals from other elements in the visual field are also processed, apparently at the same time, so that the map user can 'see' the relative positions of other features as well, and possibly transform them eventually from their size on the map to some approximation of their size in the real world. Despite the apparent simplicity of this description, the 'response' is not automatic, but depends on how the brain interprets these signals. It is clear, therefore, that the information extracted from the map is not just accepted by the map user, but is actively processed.

Obtaining information requires attention, and obviously most maps cannot be attended to all at once. It has often been pointed out that as any alert human being is constantly subjected to a vast amount of 'information' through the senses, only a limited proportion can be attended to at any one time. Therefore attention must be selective. If a map is being used as a source of information independently, then the map user's attention must be self-generated. On the other hand, maps are also used to accompany and reinforce verbal accounts, for which they can be regarded as illustrations. In such cases, the map information can be highly selective and restricted to the specific purpose of the author. Consequently, the map user's attention can be attracted to specific detail by restricting the content to a limited objective, and manipulating the design accordingly.

Attention involves the active processing of what is perceived, but this processing must be directed in some way so that what is thought to be important is attended to. Such active processing involves the ability to hold several items simultaneously in short-term or active memory, the capacity of which is limited. When map use is independent, the user must attempt to select what is pertinent, without being distracted by non-related items. When use is directed, what is pertinent be much more rigorously controlled.

Once the desired information has been obtained (in this case the location of the river), whether or not it is retained will depend on many factors, including the map user's intentions, his degree of interest, and his ability to store and recall information obtained from graphic images. If the enquiry is casual it may be forgotten in a short time. If the map user intends to learn about the location of the feature, he is likely to study the map for a longer period, and to concentrate more carefully on particular details. Although much remains to be discovered about how information is represented and stored in the brain, it is at least clear that it is meaningless to talk about map users as though they were a homogeneous body of people with identical reactions, intentions and capacities.

What takes place between the initial stimulus in the retina and the storage and processing of information in the brain is only partly understood. This is a major research field in both physiology and psychology. It is generally agreed, however, that only a small part of the information picked up by the sensory system is retained for more than a moment, and what is eventually stored is not some complete mental replica of the object, but the product of selective and highly elaborate processing.

Any attempt to explain what is involved in the use of maps must enquire into many different fields of study. At present, any such attempt at explanation will be inadequate and incomplete, for our knowledge of how the brain works is little more than a series of hypotheses. Even so, we know enough about vision and perception to suggest some of the stages in the process, and to advance some hypotheses as a basis for discussion. A full account of how information is acquired from maps must take into account the characteristics of maps as information stores, the restricted nature of their availability, the physiological and psychological processes of using them, and the cognitive processes of learning and remembering.

Examples of map-using tasks

This discussion of some of the problems involved in understanding maps inevitably leads to the introduction of many theoretical aspects of visual perception and the structure of map information. Such ideas are described and advanced in the expectation that they will contribute to a more systematic analysis of the physiological and psychological factors concerned in map use. Although some of the hypotheses can only be proposed as potential or partial explanations, such theoretical analysis has to proceed along with experimental research.

In order to facilitate this examination of the relationship between physiological, psychological and philosophical theories and the particular problems of map use, three examples of map-using operations have been selected. Although they represent only a small fraction of the total range of map uses, they should be reasonably familiar to many people, and they are intended to represent some of the variety of tasks for which maps are used. They will be referred to at many points in the ensuing chapters.

Example A

For the first example (see Fig. 1) let us assume that someone hears the name of a foreign city on a news broadcast, for instance the city of Vladivostok. Not being a geographical scholar, he has only a slight idea of where it is, but at that moment is sufficiently curious to want to find

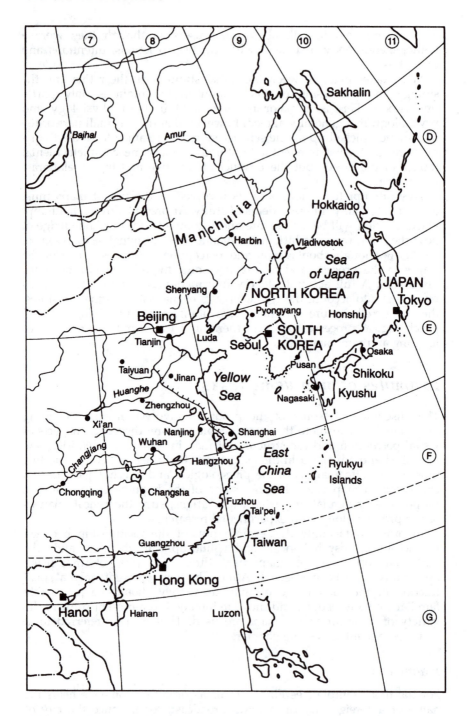

Figure 1. Map example A: a monochrome representation of part of an atlas reference map, scale 1:30 million

out. He thinks that it is somewhere in eastern Asia, but is not quite sure. He may have an old atlas in the house, or may borrow one from the children. What does he do?

What he is most unlikely to do is to carry out a random visual search. He looks up the name in the index, finds a reference to a page and a graticule unit, finds the page, and follows the reference. Within a small area of the map he searches for the name. As there is no other named town he finds it without difficulty. This is a simple, but quite common, map-using task, and it should not be dismissed because it is elementary. It should be noted that the user's intentions do not involve any deliberate learning process, and that he does not intend to 'do' anything with the information subsequently.

Example B

A motorist drives regularly between two places (see Fig. 2). She knows the route well and does not need a map to follow it. On one occasion she decides to vary the route in order to go through a small town which lies some distance from the main through route. It can be approached by various routes along minor country roads. Rather than just follow signposts she decides to look at a map and select the most interesting route from several alternatives. She finds a road map of the area, quickly locates the general area of interest (because she has used the map many times before), and begins to study the map in detail. She notes the point on the main road where she should turn off, and then concentrates on the minor roads. After comparing various alternatives she decides on one preferred route. She tries to remember the sequence of road numbers, turns and connections. On the actual journey she starts off correctly, but soon has to look at the map again to confirm that she is on the right road and heading in the intended direction.

In this case the map is being used to anticipate a future action, but is also employed to verify her position. Even though she has a good knowledge of the locality and is familiar with all the places in the area, she is unable to remember exactly all the details of the selected route. Whether or not she follows the route correctly will in due course be confirmed by her actual progress. If her reading or interpretation of the map has been at fault, then sooner or later this will become obvious on her journey.

Example C

For the third example (see Fig. 3), let us take a typical special-subject map, such as a population map of urban places in a given region. Let us also assume that the map is being used by a student, who has been directed to study it in order to learn about some aspect of the distribution of population. The map may have been prepared to accompany a text on the subject, or it may appear as a special map in an educational or reference atlas.

On this occasion, the whole of the map, and the whole of the region

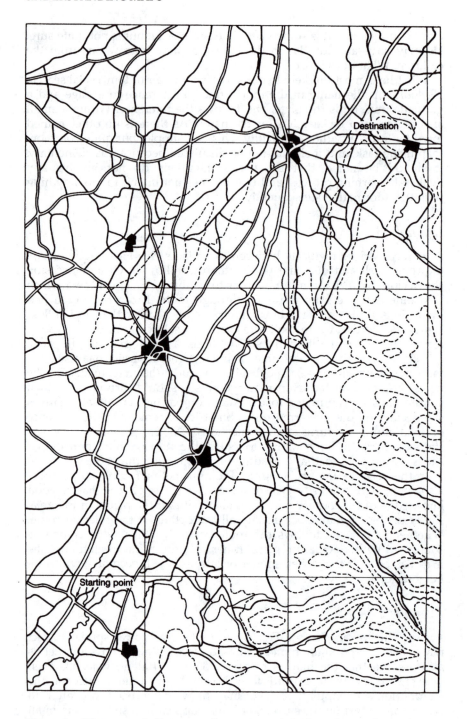

Figure 2. Map example B: a monochrome representation of part of a road map, scale 1:250 000

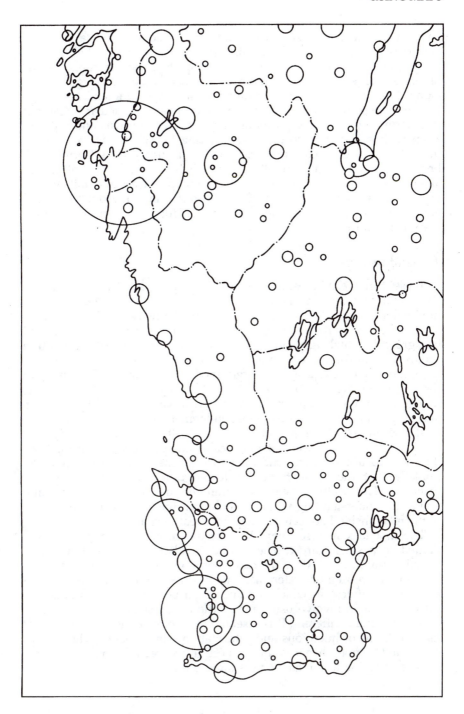

Figure 3. Map example C: a monochrome representation of part of a
population map, scale 1:2 million

concerned, must be examined, and the 'spaces' also have to be considered. The student is already familiar with the general geographical character of the region and its major features. So he observes, for example, that there is a concentration of larger towns along the coast, and widely separated small towns in inland areas. The map will be scanned many times, not to locate any particular target, but to determine the relative positions and sizes of the towns and villages. Even with prolonged scrutiny, his resulting knowledge is likely to be limited to a series of conclusions. Whether or not these are 'correct' will be very difficult for him to tell, or anyone else to verify. He will not be able to reconstruct an exact 'image' of the map, correct in detail, and an attempt to draw it from memory would result in many omissions and errors. In writing an essay upon a particular topic he might refer to the map many times.

Types of map use

If these three examples are analysed in general terms, it can be concluded that the first task involved finding a named feature, with minimal pre-existing knowledge; the second involved the study of a limited area to solve a problem posed by the map user, with good background knowledge and a clear intention of anticipating a future action. The third required the examination of a whole map, or part of a map, the extent of which was determined by the subject of study, with the deliberate intention of learning about a subject. In the first example, the scale of the map was probably regarded as immaterial by the user, and he probably made no conscious judgement about scale or distance. In the second, the user would find it difficult to select or anticipate a route without being aware of map scale, and possibly converting map distances to approximate true distances. In the third case, it is again unlikely that the student would devote much attention to specific distances. The relative distances within the area would be important, and a full understanding would assume that the student had some idea of the relationship between the map and 'real' distances.

The first task required very little visual search, and even that was limited to a single target. The second involved a limited but highly selective search directed by the map user's own judgement of what was relevant, interesting or preferable. The third required extensive search in unfamiliar patterns in which nothing could be neglected. Although all three operations involve using maps, the circumstances and intentions of each are quite different. It is essential to recognize the wide variety of uses, in terms of intentions and information requirements, which maps cover, and to note that it is the connection between the map and what the user is trying to do that is most important.

CHAPTER 2

An outline of the visual system

The three examples of map types given in the first chapter are by no means comprehensive, but should be enough to serve as illustrations of the fundamental processes involved in using maps. It is necessary to consider what elements they have in common, and how they are specifically different. Such an analysis requires the breakdown of map use into a series of activities.

Whatever task is concerned, the initial stage depends on the two processes of detection and discrimination. At the most basic level, the map user must be able to respond to what is there; that is, the symbols on the map must be a sufficient stimulus to make them detectable. In addition, he must be able to distinguish between one symbol and another, and therefore discriminate between them. To some extent this will depend on his ability to look carefully, and in some cases to notice small differences. In most types of map use, these processes will operate subconsciously during identification, in which the symbols in the area of concern are perceived and understood.

Identification

These three aspects – detection, discrimination and identification – are often regarded as equivalent to 'map reading' in the simple sense; that is, the perception and ready identification of the symbols on the map (see Fig. 4). It will be argued in due course that this is not the equivalent of 'reading' a text. Whereas detection and discrimination can take place without understanding what the symbols represent, identification is clearly learned behaviour. For a familiar map, and one much used in practice, it may not be necessary to consult the legend at all, in which case the symbols are identified by being matched against some representation held in long-term memory. On the other hand, a person using a complex geological map for the first time might have difficulty in remembering the large number of area symbols, and often fail to identify correctly the map symbols without frequent reference to the legend.

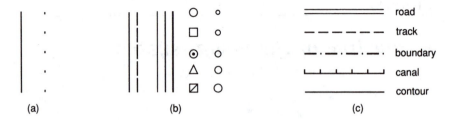

Figure 4. (a) detection, (b) discrimination, and (c) identification

Identification and discrimination are closely connected. Although in most cases the differences between two symbols (for example, a continuous brown line and an interrupted red line) will be readily apparent, to the point where perception seems to be instantaneous, there are occasions when errors in identification will occur because of errors in discrimination. An interrupted black line representing a footpath may be superficially similar to one representing a boundary. Failure to respond to small differences may cause confusion in identification. If map users can appreciate the graphic variables of form, dimension and colour which are systematically employed in map design, there is a greater likelihood that they will be aware of, and therefore respond to, small amounts of visual contrast. Although discrimination depends in the first place on the degree of contrast between symbols created by the map maker, it also requires an ability on the part of the map user to respond. This can be affected by practice.

Recognition

From the point of view of map use, identification should be clearly distinguished from recognition (see Fig. 5). In psychological and physiological literature, the two terms are often used as alternatives, but Forgus and Melamed (1976: 18) make the distinction clearly: 'Recognition means being able to say that something looks familiar, whereas identification means that we can say what it is or name it'. Although both terms imply familiarity, it is an advantage to distinguish between identification of a symbol and recognition of a geographical feature. A blue line may be identified as a shoreline, and a pale blue colour as representing an area of water. But some geographical features, such as the Great Lakes of North America, or the Mediterranean Sea, are likely to be recognized as such because their shapes are familiar. Recognition of geographical features is important in learning, and indeed new spatial information is often 'placed' by being located in reference to familiar features. The recognition of a particular geographical shape, despite the variations introduced by changes in scale, projection, orientation and generalization, raises the interesting question of how such information is composed and stored in the brain.

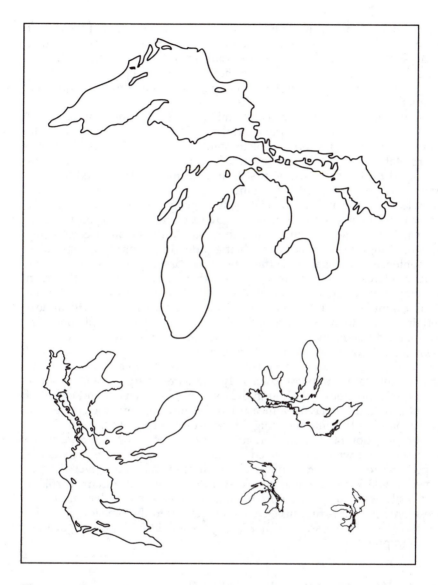

Figure 5. Recognition

Interpretation

Detection, discrimination and identification can be regarded as the preconditions for map use. Beyond this point there must be a further stage of interpretation, by which the information is processed in order to be actively employed to deal with a particular map-using task. It would be convenient if it was possible to make a clear distinction between the identification and interpretation stages, but this is not so. The understanding of the meaning of a map symbol is an interpretative act, because of the nature of map symbols, and the process of looking at the map is itself guided by the intentions of the map user. The isolation of these stages or 'processes' is necessary in order to be able to discuss the subject matter verbally. But this should not be accepted as indicating that the processes themselves operate in some simple sequential manner.

How visual information is assimilated, stored and subsequently recalled and used remains the central problem in visual perception, to which there is as yet no complete solution. However, it is clear that interpretation always goes beyond identifying the individual items displayed on the map. In the first example, finding the named town in Asia may be followed by some attempt to assimilate the new knowledge; for example, by relating it to geographical features already known and therefore 'recognized', or by reaching a conclusion verbally, such as 'it is nearer to China than I thought'. In the second example, the various minor roads selected are being judged not simply for speed or convenience, but against certain personal assumptions about scenic beauty, interesting views, etc., for which there may be some evidence on the map, but which will also be affected by previous experience. In the third example, the interpretation of the population map will almost certainly lead to a comparison with other geographical phenomena. Although different map users may produce interpretations which have elements in common, they will also vary with individual skill and knowledge. There can be no guarantee that two people using the same area of the same map for ostensibly the same purpose will produce similar interpretations. Although the total number, density and extent of symbols on a map may be a measure of its 'information', the conversion of any or all of this into meaning depends on interpretation.

Interpretation raises many interesting problems. How are connections made between what is perceived on the map, and other knowledge? Is it possible to 'learn' from maps about things which are never directly experienced? Why do some people seem to learn to use maps correctly and easily, while others do not? To begin to have some understanding of these problems, it is first necessary to enquire into the physiological and psychological processes which underly visual information acquisition and interpretation.

An outline of the visual system

The visual system comprises a receptor system, the eyes, linked to a processing system, the brain. A schematic diagram of the basic arrangement is shown in Fig. 6. Although the lens and ciliary muscles of each eye are important in controlling focus and direction, to some extent defects in this part of the optical system can be corrected or minimized. The record of the pattern of light entering the eye is initially formed in the retina (Fig. 7), a complex arrangement of cells, some of which are light-sensitive or photoreceptors. The retina is capable of adapting to the general degree of illumination, and within this overall range responds

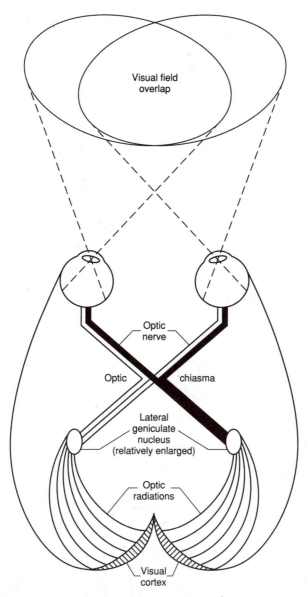

Figure 6. Schematic diagram of the visual system

to small differences in intensity of radiation from different parts of the visual field. The effect of light rays entering the eye is to bring about a chemical change in the retinal cells affected, which in turn produce minute electrical discharges. These discharges are the 'signals' which are eventually transmitted through the ganglion cells into the optic fibre tract, which is linked via the lateral geniculate nucleus to the visual cortex. It would be very simple if there was a one-to-one corres-

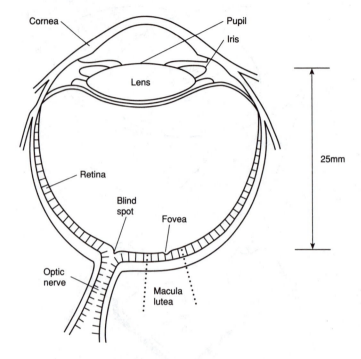

Figure 7. Schematic diagram of the eye

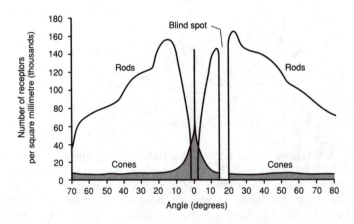

Figure 8. Relative distribution of rods and cones in the retina

pondence between the retinal cells and those in the visual cortex, and if all
the retinal cells behaved in an 'on' or 'off' manner, but this is not the case.
The retinal cells are of different types, they are unevenly distributed,
and they are interconnected. Consequently, the retinal 'image' which
results from viewing the visual field is not a copy or duplicate of the
external stimulus, but a selectively processed version of it.

The retina

There are two types of photoreceptors: rods and cones. The rods will
respond to small changes in intensity, but are insensitive to differences
in wavelength. The cones need a greater degree of illumination in order
to react, but they are also sensitive to differences in wavelength. A small
area of less than 1 mm in diameter in the centre of the retina (called the
macula) is tightly packed with cones, whereas the periphery of the
retina contains mainly rods. The cones in the macula are thinner than
elsewhere (only 1.5 μm), and in the middle of the macula the area
known as the fovea has all other nerve cells and layers pulled aside to
expose it directly to light entering the eye. The relative distribution of
rods and cones is shown in Fig. 8.

Each retina has about 120 million rods and about five million cones,
but there are less than one million optic nerve fibres leading from the
ganglion cells. In the outer periphery of the retina, as many as 600 rods
are connected to one optic nerve fibre, whereas in the fovea there is an
almost one-to-one connection between cones and fibres. This helps to
explain why maximum visual activity occurs in the fovea.

The rods and cones are photoreceptors. The effect of light, or of a
change in light intensity above threshold, is to decompose
photopigments, which in turn causes an electrical signal, which is
eventually transmitted to a ganglion cell. Between the photoreceptors
and the ganglion cells there are three other cell types: the bipolar,
amacrine, and horizontal cells (see Fig. 9). The bipolars receive and
transmit signals usually from several receptor cells to one ganglion, but
in the fovea there is usually one for each cone. The horizontal cells are
connected to adjacent bipolar cells; and the amacrine cells have a rapid
but very brief response to any change in stimulation, their function
being to link ganglion cells.

The consequence of these cell connections is that each ganglion does
not simply transmit an impulse from a single point on the retina, but
summarizes the signals received from a number of inputs. Therefore,
each ganglion cell has a receptive field, and the receptive fields of
ganglion cells overlap. To see the effect of this in the visual system, it is
necessary to digress briefly to consider some basic facts about neurons
and their interconnections.

Neuronal activity

In a typical multi-polar neuron there is a cell body (soma), extending
into a dendron, which consists of a series of branches called dendrites

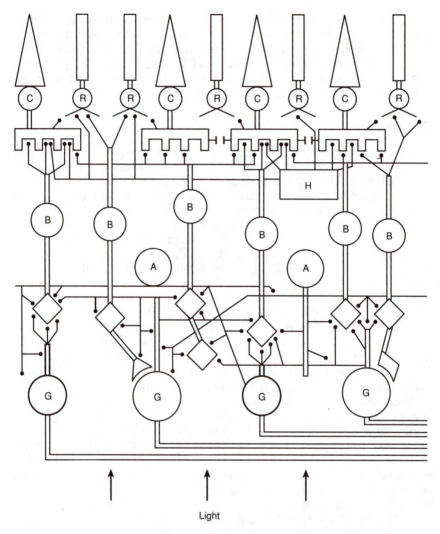

Figure 9. Schematic diagram of retinal cells and interconnections. C=cone; R=rod; H=horizontal; B=bipolar; A=amacrine; G=ganglion

(see Fig. 10). It is these which form the principal connections with other neurons to receive incoming signals. The axon extends from the cell body, and signals are transmitted to other cells through the axon terminal. The point at which an axon terminal approaches the dendrite of another cell is called a synapse, although a synapse can also occur directly on a cell body.

The interior of any such cell is negatively charged in its resting or passive state, and this is known as its resting potential. Incoming signals may bring about a change in potential. It may become less negative, in which case it is depolarized, or more negative, in which case it is

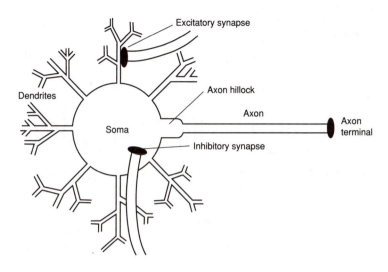

Figure 10. Schematic diagram of multi-polar cell, showing synapses with axon terminals of other cells

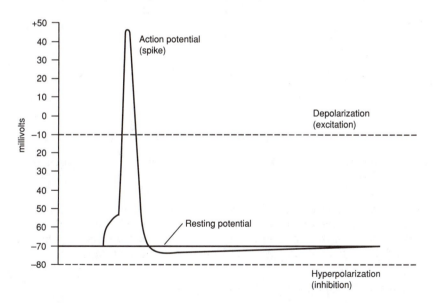

Figure 11. The action potential (spike) of a discharging neuron

hyperpolarized. This means that a cell may respond by becoming either excited or inhibited. For a cell to discharge a signal, the incoming stimulus must exceed a certain threshold to produce an action potential (Fig. 11). The intensity of the stimulus is reflected by the rate at which the cell discharges. These discharges, which are all of the same strength,

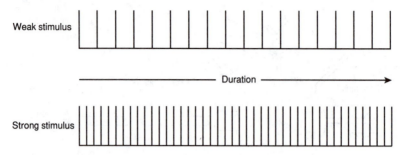

Figure 12. Stimulus intensity and neural response: 'spike' frequency

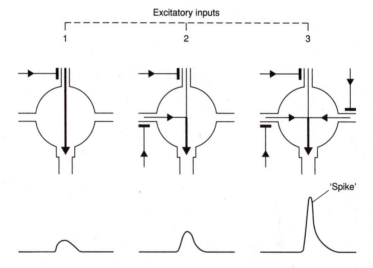

Figure 13. Summation of excitatory inputs causing a neuron to reach discharge level

are commonly referred to as 'spikes', because of their appearance on a graph (see Fig. 12). As each neuron is normally receiving signals from many other cells, the final output is determined by summation, that is, the combined effect of excitatory and inhibitory inputs (Fig. 13).

Ganglion cell responses

In consequence, the basic pattern of inter-cell connections is that many (sometimes thousands) of synapses converge to form the total input, and the cell's discharge may diverge through many connections to the dendrites of other cells. The structure of different cell types is of course far more complex than the simplified version represented by this outline.

Stimulation of a point on the retina can lead basically to three types of response in a ganglion cell. It may produce a definite and rapid

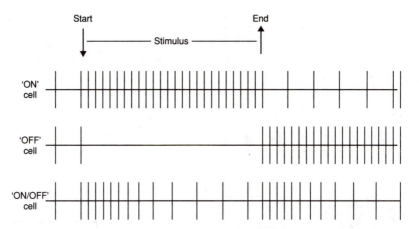

Figure 14. Types of neural response: 'on', 'off' and 'on-off'

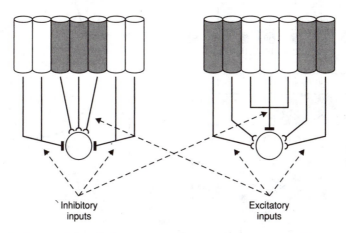

Figure 15. Section through the receptive field of an 'on-centre' cell and an 'off-centre' cell

discharge (excitation) which persists as long as the stimulus lasts. This is usually referred to as an 'on' response. It may decrease the activity of the cell during the stimulus, but then respond with a brief burst of activity when the stimulus is terminated. This is known as an 'off' response. And third, a brief burst of activity may take place at both the onset and the termination of the stimulus (that is the point at which the relative intensity changes), and this is called an 'on–off' response (Fig. 14).

The combination of all three types is typical of ganglion cell receptive fields. In some cases the centre of the receptive field gives an 'on' signal, whereas the periphery responds with an 'off' signal. The margin between the two has an 'on–off' response. It seems that the converse pattern with an 'off-centre' field is equally common (Fig. 15).

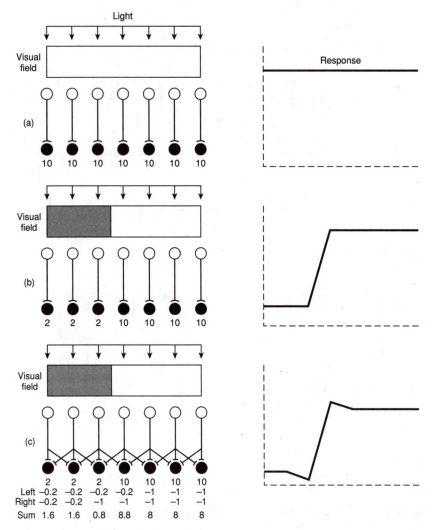

Figure 16. Lateral inhibition along an edge

Lateral inhibition

The effect of these different reactions in the receptive field is of great importance in the visual system, for it accounts for an increase in apparent contrast along edges (brightness differences in the visual field). This effect is called lateral inhibition, and its operation is exemplified by Fig. 16 (based on Paul, 1975: 104). At (a) the receptors are equally stimulated and therefore excite each individual neuron to an equal degree. The resultant discharge from each neuron is represented arbitrarily by 10. In (b) the illumination is reduced over part of the visual field, so that the excitation of some neurons is reduced proportionately to 2. In (c) the neurons are laterally interconnected, the adjacent connections being inhibitory. If the inhibition is assumed to be 10 per cent of the excitation, the cumulative effect is both to decrease and increase the response along the edge. The discharge pattern of the neurons is shown diagrammatically on the right.

Visual processing in the brain

The optic fibres from the temporal and nasal halves of each retina converge at the optic chiasma (see Fig. 6), where they divide in such a manner that the information from one side of each eye is directed to the same side of the visual cortex. Because the retinal image is reversed, this means that the left side of the cortex deals with information from the right side of the visual field and vice versa. Binocular stereoscopic vision is of course vitally important as a general visual function, because it permits the perception of depth and distance, but it is not important in viewing the limited two-dimensional plane of the map.

From the optic chiasma the optic fibres terminate in the lateral geniculate bodies, and from these optic radiations extend into the primary areas of the visual cortex. The cell groups in the cortex have further connections with the secondary visual association areas, which in turn lead to the general interpretative area at the centre of the brain.

The six layers of the lateral geniculate nucleus are divided equally between the information supplied from each retina, and therefore contain the first stage in stereoscopic vision. The cell arrangements in each lateral geniculate nucleus are similar to those of the ganglion cells of each retina, but far more are devoted to contrast and movement, and fewer to luminosity. Similarly the paired responses for red–green and yellow-blue are also found in the lateral geniculate nucleus, suggesting that its main function is to carry out further processing of incoming signals.

The cells of the visual cortex also have a topographic relationship with the ganglion receptive fields of the eyes. About 50 per cent of each half of the cells is devoted to the macula, with the fields from other retinal areas being represented in their correct spatial relationship.

The primary visual cortex (area 17) projects further into areas 18 and 19 of the brain, which are usually known as the visual association areas (Fig. 17). Although similar in arrangement to the cells of the primary cortex, they appear to deal with more complex processing, in particular, complex shapes and arrangements. Some of the evidence for this has been derived from people with brain damage in these areas, whose ability to identify or recognize shapes has been destroyed or severely impaired.

Although the lateral geniculate nucleus appears to concentrate on the contrast at edges and the discrimination of form, the primary visual cortex carries the process of form analysis even further. It appears that the cortex contains columns of cells, extending through six layers, and

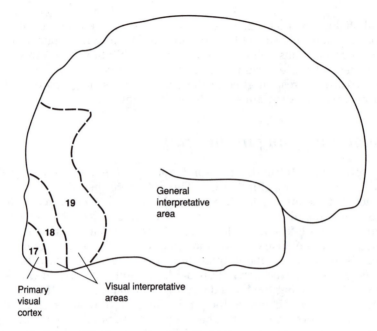

Figure 17. Areas of the brain primarily devoted to visual perception

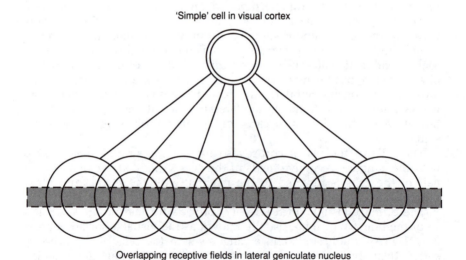

Figure 18. Schematic diagram of an array of 'on-centre' receptive fields in the lateral geniculate nucleus, forming a linear receptive field (line detector) for a 'simple' cell in the visual cortex

these are specialized to detect particular visual units – line widths, angles and so on – which seem to be components for the discrimination of shape and pattern. Others respond only to objects moving in particular directions. Some complex cortical cells have typically large receptive fields, and seem to respond mainly to movement in the visual field, whereas the receptive fields of simple cortical cells appear to be specialized for lines, edges or bars at particular angles (see Fig. 18). At an even higher level, hypercomplex cells appear to respond only to specific line widths, lengths and angles. The existence of such fields was first discovered by Hubel and Wiesel (1968) in the cortex of the monkey, and they have since been postulated in the human brain.

The overall impression given by recent physiological research is that of a sequence of processes, capable of initially detecting edges, lines and patterns, and subsequently analysing these into more complex structures. It can be inferred that a regular geometrical feature, such as a square, is analysed as an outline consisting of two parallel vertical lines and two parallel horizontal lines. Presumably the analysis of an irregular shape, such as a geographical feature, requires longer and more complex processing, more connections, and therefore is more difficult to 'learn'. In this sense, any visual scene or image may be broken down into shapes or patterns, and the brain is then able to operate on them, presumably being able to compose or regroup them as a process of thought or imagination.

These concepts of course only attempt to describe what happens, and to distinguish those cells in the cortex which appear to control specific mechanisms.

This does not explain how these processes operate, or how the brain converts such analytical patterns into meaning. It also raises a fundamental question in human perception: is visual activity a set of processes which start with a stimulus in the receptors and then proceed in some order to higher levels (data driven); or does the whole process depend primarily on a chain of events initiated by the brain, which directs the perceptual system towards certain goals (concept driven)? Both Oatley (1978) and Young (1978) suggest that the flexibility of the human brain is such that probably both can be operated, depending upon the circumstances and the nature of the problem involved.

Oatley (1978) points out that the rather low sensitivity of the visual system to both high and low frequencies is counterbalanced by its ability to adapt locally to brightness differences (contrast). This of course is a function of lateral inhibition, and Oatley suggests that a receptive field of the 'on–off' centre-surround type can be regarded as a frequency filter. The hypothesis, therefore, is that one function of the visual system is to undertake frequency analysis. As Oatley (1978: 205) puts it

The image in the retina suffers severe degradation due to a number of optical and neuronal factors. However, the visual system because of the organisation of receptive fields is able to separate different spatial frequencies in the image into different channels. This process may be thought of as similar to making a Fourier transform of the image. The advantage of operating in this domain is that if the degradation characteristic of the system is known, it is a relatively easy matter to adjust the gain of each channel.

The limitations of contrast sensitivity in the human visual system are of course exploited in the construction of maps. The fine screens of half-tone and tint are deliberately kept below the resolution threshold. The highly disturbing effect of broad parallel lines at a given frequency in some patterns is well known, and therefore consciously avoided. The deliberate enhancement of contrast between area symbols which differ little in brightness and wavelength characteristics by dividing them with a higher contrast line is also a familiar practice. On the other hand, the use of very high contrast in some parts of a map compared with others is also visually disturbing, and this may be due to the need for local adaptation.

All cartographers are aware of the importance of contrast in map design. A better understanding of the contrast sensitivity of the human visual system would be extremely valuable if the principles involved could be applied systematically in the context of particular maps.

Eye movements

As most maps contain comparatively small and detailed images, both detection and discrimination are likely to require foveal (or central) vision. Assuming that focusing is no problem, and that the map is being viewed at the correct distance, the visual field of the map can only be examined in detail by looking at individual bits of it in a series of movements. Although several types of eye movements are involved in vision (including conjugate movements to direct both eyes together; tracking or pursuit movements to follow a target, etc.), the most important ones in the visual perception of a map are the saccadic movements which connect eye fixations. For each fixation, the eyes are 'fixed' on a certain area of the visual field, and they are successively moved to bring different small areas into central vision. These small eye movements are supported by larger movements which 'jump' from one area to another, and by head movements which enable a larger area to be scanned. Fixations vary in duration, depending upon the complexity of the target and the time taken to perceive it, but generally the eyes will drift away after a few seconds, and will have to be deliberately redirected if the fixation is to be sustained.

Fixations can be both voluntary and involuntary, and they are controlled by different parts of the brain. In addition to drift, the fixation is subject to a continuous fine tremor, due to tension of the eye muscles, so that a fixated target is always slightly 'moving' on the retina. Ditchburn (1973) points out that a stationary target is never completely fixed in one place on the retina. The effect of this on the retinal photoreceptors is by no means fully understood.

Therefore, when a map is being looked at to examine specific details, information is acquired through a series of separate steps. Although the whole visual field may be perceived simultaneously, a series of fixations on areas of detail necessarily implies that the information is taken in and stored, at least for a short time, so that it can be 'put together'. In reference to viewing pictures, Kolers (1973: 25) observes that 'not all the information in a picture is apprehended at once; rather, the information in a picture is processed in time'. To take an obvious example, if an unfamiliar symbol is perceived on a map, it is fixated while it is being examined for colour, form and dimension, and then the next fixation will probably be similar symbols in the legend. Several symbols may be scrutinized, to see which one matches the stored image. It is clear that the visual system is so arranged that although the whole visual field can be seen in a broad sense, the inspection of fine details (which puts a much heavier burden on the information-processing operation) is carried out a bit at a time.

The sequence of fixations recorded in viewing a particular map has been analysed by Castner and Lywood (1978) and Castner (1979a). As might be expected, series of fixations concentrated on areas of contrast are linked by larger 'jumps' to other areas but the actual patterns of eye movements are unique to the individual. An interesting test by Chang, Antes and Lenzen (1985) showed that experienced subjects used fewer and shorter fixations to carry out an interpretation of a topographic map than inexperienced subjects. This confirms that the skill of the attentive map user is built up by practice and experience.

Memory

Despite the high rate of normal saccadic eye movements (from a few milliseconds to one or two seconds) it is clear that a series of retinal records acquired through separate fixations must be stored, as each retinal image is rapidly replaced by another. In map use, all three types of memory are involved. Brief visual, iconic or sensory memory is the name given to the storage of a fixated image long enough for it to be processed by the visual system. Short-term memory, usually available for a few seconds to a few minutes, can store and operate on a limited number of items simultaneously. And long-term memory can store an enormous number of items, but consequently provides problems in recall.

Physiological explanations of memory at present are based on the reverberatory circuit theory, which suggests that a neuronal circuit brought into action by a given stimulus continues to repeat the signal for a short time. Repeated or deliberate rehearsal can consolidate this into long-term memory. On the basis of recent research it is known that anatomical changes take place which establish permanent neuronal circuits, thus allowing signals to pass with greater facility. The effect is most marked at synapses, and Blakemore (1977: 114) refers to 'modifiable synapses', based on the investigations of Young. The more often the same circuit is used, the greater the level of consolidation.

It has also been suggested that information obtained from iconic or brief visual storage can pass directly into long-term memory, if the experience is sufficiently vivid. This would suggest that a visual image, or some aspect of it, can be stored directly in long-term memory. The connections between vision and memory are clearly important in studying questions of map use.

The circuits produced through this linking of many neurons are now regarded as of primary importance in understanding how the brain works. On this basis, it is accepted that memory is not some separate 'store' which can be accessed, but that the established neuronal circuits are themselves memory. This characteristic of forming neuronal circuits is believed by many researchers to be the prime force in learning, and the way in which the human being develops and adapts to the external world. Significantly, it is the study of neuronal circuits as the physiological basis of knowledge and learning which has led to the concept of the 'learning' computer. This will be taken up in Chapter 20.

CHAPTER 3

Visual perception and map interpretation

Detection

The detection of an image in the visual field is primarily dependent upon two sets of factors: the optical system of the eye, and the receptor system of the retina. These have been likened to the lens and the photographic film of a camera. The lens has to focus the image on the retina, the image being reversed laterally and inverted. The convex lens can be changed in shape, so that objects at different distances can be brought into focus.

Visual acuity therefore depends in the first place on the efficient functioning of the lens system of the eye, but this is only the starting point. Thereafter, it is controlled by how the retinal cells react to the incoming radiation. Visual acuity is usually expressed as the minimum object size that can be detected at a given distance under certain conditions of contrast, and this is often determined by viewing a test pattern containing fine lines or shapes such as letters. However, it is also evident that the resolution of fine detail should also depend on the cones in the fovea, these being the smallest individual units which can be stimulated. A high level of contrast is also assumed.

Optically, the controlling factor is the angle subtended in the eye by a given object at a specific distance. If the object being viewed is moved further away, then it must be made proportionately larger to subtend the same angle. However, whether or not it will be detected will also depend on what constitutes a sufficient signal for the photoreceptors in the retina. In the foveal region, the cones are about 1.5 μm in diameter. It would seem that adjacent small points could be detected as separate if in the retinal image they were at least 2 μm apart on the retina. As the fovea is less than 1 mm in extent, maximum visual acuity is limited to a very small area, some 3° of the total visual field. In the periphery of the retina, where many rods are eventually connected to one ganglion, a larger area of illumination (although at a low intensity) is needed to cause any reaction, and therefore acuity is lower. In considering the importance of foveal vision, Osaka (1980: 45) points out that 'information necessary for semantic identification is obtained from the foveal and near parafoveal region.'

It has already been pointed out that the resolution threshold for the human visual system is quite low. This is presumably a function of human evolution, as visual acuity to hundredths of a millimetre is not critical for the normal purposes of everyday life. However, visual acuity in this sense is only one aspect of the process of detection. It is also necessary to be able to locate objects, that is, to detect the relative position of objects or patterns. The previous discussion has demonstrated that the ganglion cells have interconnected receptive fields which also overlap. As Horridge (1968: 217) puts it, 'The great problem in the sensory system is that although behaviour leads us to expect evidence of discrimination of central cells, the records from units show mostly summaries.'

However, if a particular small image occurs at the overlap of three receptive fields, then it can be localized to fine limits. Leibovic (1972: 273) explains this as follows:

> The loss of spatial resolution due to the extent of a receptive field can be recovered in the responses from overlapping fields. In fact, acuity in the perception of fine detail can, in principle, be made better than the fine grain of the receptor mosaic in a sensory surface: it is well known that human visual acuity in distinguishing two separate stimuli is better than the separation of two foveal cones.

In the use of maps, visual acuity in this sense is not a major problem. It is highly unlikely that any map will be constructed with symbols that are too small to be seen, that is, which are below the absolute threshold of detection. This is not to say that detection problems do not occur; simply that they are likely to occur for other reasons.

The degree of contrast between a map symbol and its background has a critical effect on its detection. At the map-making stage, contrast can be reduced by making a symbol lighter, by making the background darker, or failing to anticipate an overall reduction in the illumination during map use. For example, a black line of 0.1 mm gauge can be detected quite easily against a white background. A yellow line of the same width can hardly be perceived at all. In the same way, it can be difficult to detect a blue line against a green background. In many multicolour maps, point line and area symbols occur in a great variety of combinations and juxtapositions. Problems in detection ought to be anticipated by the map maker, but it is always possible that some symbol combination will affect contrast, and therefore pose a problem in detection.

Detection problems may also arise in the process of visual search, as a map is scanned, bringing different parts into central vision. The way in which fixations are ordered is by no means fully understood, but it is generally believed that features detected in peripheral vision are likely to influence the sequence of fixations. Map symbols, especially small point symbols, close to the limits of visual acuity, may be inadequate as a stimulus for peripheral vision, and therefore may be 'missed' entirely, or found only after prolonged searching.

Symbols which are close to the limits of resolution are obviously a disadvantage if the map is to be used within constraints of time, or under pressure. Where decisions may have to be taken quickly, as for example in visual air navigation, or on a battlefield, it is imperative that everything on the map can be detected without hesitation or prolonged scrutiny.

For most maps, the minimum symbol size and contrast are judged on the assumption that the map will be used at the 'normal reading distance' of about 30 cm. If the map is viewed at a greater distance, for example by displaying it on a wall, then this may no longer hold. Maps which are intended to be viewed from a distance have to have the symbols enlarged correspondingly to remain within the limits of normal visual acuity.

Discrimination

The process of discrimination is much more critical in map use than detection, although it is possible to confuse one with the other. To discriminate could also be described as the ability to 'detect a difference'. Common requirements in map use are to perceive that two symbols, although similar, are slightly different; or that although in different surroundings, two symbols are the same. For identification it is necessary, generally speaking, to compare the symbol found on the map with that shown in the legend, or with a visual image of the legend symbol stored in memory. Compared with detection, discrimination is a complex process, and many maps exist in which it is not always possible to carry out discrimination with confidence.

Discrimination has been defined by Fellows (1968: 1) as 'the process by which an organism responds to differences between stimuli'. It is affected by both physiological and psychological factors.

Physiological factors in discrimination

To discuss physiological factors it is necessary to consider the physical attributes of a stimulus which is represented as a pattern in the retina. A perceptible stimulus has basically three attributes: intensity, extent and duration. For a normal map, whether printed or displayed, duration can largely be ignored, in the sense that the stimulus will continue for as long as the map user observes it. The extent of a stimulus can be expressed as its dimension and form. Any perceptible symbol which can be discriminated from others must have both. Its intensity is a function of its absolute luminance and its wavelength composition, which can be described broadly as its brightness and hue. A printed map viewed by reflected light has a limited range of brightness, and in any event the visual system is organized to respond to brightness differences (contrast) rather than absolute brightness. The wavelength composition of light reflected from paper or transmitted through a visual display results in the discriminable elements of hue, saturation and lightness – the properties of colour.

The minimum conditions for discrimination are sufficient difference in form, dimension and contrast between two symbols. The contrast in intensity between image and background may be achromatic, and therefore 'colourless', or it can also include differences in hue, saturation and lightness in coloured images. A small grey square, with the same form and dimensions as a small black square, can be distinguished because it has a higher lightness value and therefore a lower contrast with the background, even though it is an achromatic image. On the other hand, a small red square and a small black square of the same form and dimensions can be distinguished because the contrast with the white background also includes a difference in hue. The same symbols appearing in red and pink could be distinguished because of the difference in saturation. These elements provide the range of visual clues that can be employed to create perceptible differences, and therefore permit discrimination (Fig. 19).

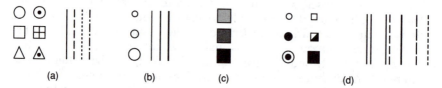

(a)   (b)   (c)   (d)

Figure 19. Some graphic variables and discrimination: (a) form; (b) dimension; (c) lightness (value); and (d) in combination

Maximum discrimination in form, dimension and colour depends on foveal vision, when the images are simultaneously present in the visual field. This can occur only in viewing a small area of the map in one fixation. When discrimination is applied to two or more images which do not appear in one fixation, then other factors, including memory, come into play.

In terms of map use, the whole image consists of an organization of points, lines and areas. Point and line symbols must have specific form and dimensions. The external form of an areal symbol depends upon the actual areal extent and 'shape' of the particular feature. Therefore its form can only be described in terms of its internal variations. If it contains perceptible differences within its edge or outline, then it has a pattern. An area symbol can be achromatic, consisting of some combination of points, lines and spaces, but if it also has colour then variations in hue, lightness and saturation can reinforce the pattern, leading to a very large number of perceptible differences.

For example, an area of water may be represented by being symbolized in light blue. In this case there is no variation internally in form or dimension, but only the contrast between the hue and the adjacent areas, which themselves may be white or coloured. On the other hand, it may be represented by a pattern of blue lines, in which case the colour distinction is reinforced by internal variations in form and dimension, that is, the continuity, thickness and spacing of the lines in the pattern.

Physiological factors, therefore, apply mainly to the characteristics of the stimulus, and the ability of the visual system to discriminate between different aspects of the stimulus.

Psychological factors in discrimination

Discrimination and attention

Generally speaking, a map can be perceived at two levels. Compared with the total extent of the visual field in normal stereoscopic vision, the map image is relatively small. Therefore, maps can be 'seen' in their entirety, using both peripheral and central vision. Initially only the broad forms and most salient features will be noticed. This level of attention may well be unstructured or involuntary, and governed largely by the accident of the spatial arrangement of the most prominent elements, the presence of large or distinctive shapes, or recognizable features. For the extraction of specific information in a map-using task, discrimination requires the direction of attention, and therefore the selection of what is judged to be important. Although selection has already taken place in the map content, this can only subserve broad objectives of map use. The particular details in a map which must be noticed and interpreted for a specific task are likely to constitute only part of the total map content. Short of constructing a map for only one task, the map cannot be designed in advance to provide the ideal selection of information for a particular map-using operation.

This aspect of map use is of course analogous to other forms of information acquisition through the senses. Many psychologists have pointed out that a human being ordinarily attends to only a fraction of the sensory information available. When a particular problem or requirement is being dealt with, then attention is directed to those aspects of the information which appear to be relevant. As Navon (1977: 355) puts it, 'In most real situations the task of the human perceptual processor is not just to account for a given input but also to select which part of the surrounding stimulation is worth receiving, attending to, and processing.'

With a simple, finite task, like finding a given location, attention can often be directed to a narrow field, and the non-relevant items ignored. Where more complex interpretation is necessary, in which the map user quite often cannot decide what information is pertinent until after she has examined it, then both broad spatial arrangement and specific detail have to be attended to. As Polyani (1969: 125) says, 'Every time we concentrate our attention on the particulars of a comprehensive entity, our sense of its coherent existence is temporarily weakened; and every time we move in the opposite direction towards a fuller awareness of the whole, the particulars tend to become submerged in the whole.'

In some cases, what is being sought in the map will firmly control the elements selected for attention. On the other hand, previously unnoticed features may demand attention as they become relevant to the task. Brown (1958: 24) makes the point when he gives the example that 'It is clear, however, that when a dirt road turns into a four-lane super-highway, we may for the first time notice the difference between a thin and a thick red line on an actual road map.' Close attention will lead to the discrimination of details, and in turn discrimination will itself depend on a sufficient level of attention.

Psychophysical factors in discrimination

The relationship between a stimulus (which must be some form of external physical energy) and a sensation (the response which results) constitutes the field of study of classical psychophysics. From the early nineteenth century, attempts were made to formulate rules, or 'laws', which would state systematically the relationship between stimulus and sensation. The main enquiries were directed towards establishing the minimum stimulus which could be detected (absolute threshold), and the just noticeable difference between two stimuli (difference threshold).

The problem as seen by the psychophysicists was clearly expressed by Stevens (1975: 3): 'The issue concerns the form of the psychophysical law, the equation that tells how the strength of the external stimulus determines your impression of subjective intensity.' The equation had been pursued by Stevens in many attempts to find the particular ratios which fitted the different senses. This resulted in the so-called power law,

$$S = KI^y$$

in which the sensation (S) is stated to be equal to a constant (K) times the intensity (I) of the stimulus, raised by the exponent (y). As he puts it (Stevens, 1975: 16), 'A general rule is this: on all continua governed by the power law, a constant percentage change in the stimulus produces a constant percentage change in the sensed effect.'

But the relevance of Steven's theories, and the power law in particular, is very much dependent upon the aspect of the stimulus being measured. Most of his experiments, and the most convincing results, are concerned with stimulus intensity, e.g. brightness, loudness. But most of the discriminable differences between map symbols are not simply a matter of intensity alone, and consequently the relevance of Steven's theories to the perception of map symbols cannot be automatically assumed. Attempts to do so illustrate the dangers of importing psychophysical theories into considerations of map perception without a careful scrutiny of what aspect is being measured or compared.

Sensitivity

Psychophysical experiments are concerned mainly with sensitivity of the nervous system to different degrees of stimulation. It has been pointed out already that a physical stimulus can be described in terms of intensity, extent and duration.

In investigating the responses of human observers to a graphic image, these differences must be clearly understood. In viewing a stationary two-dimensional image, the duration of the stimulus will normally depend upon the time needed to comprehend it. The image is perceived through light reflected or transmitted, and in a printed image this is reduced by the pigments of the printing inks, which act as filters. As the map will normally be used in daylight, or adequate artificial light, some will be absorbed by the paper surface, and the maximum intensity of the light reflected will be quite low. Although overall brightness can be varied in most visual display units, the maximum brightest level has to be strictly limited for ease of viewing. Therefore, differences in intensity are limited to quite a narrow band, if compared with the total range which the visual system can accommodate (over one million to one). Discriminatory responses to map symbols, therefore, depend more on contrasts in form, dimension and colour, than on intensity alone. Differences in intensity are most important in establishing threshold levels for contrast, form, dimension and colour.

Many psychophysical experiments have been devoted to the comparison of intensities, for example by observers viewing projected spots or beams of light, or holding different weights. Stevens (1975: Table 1), lists a large number of these. But on a map, a large blue area has the same intensity of stimulus as a slightly smaller blue area. They differ in spatial extent, not intensity. Contrasts along edges, such as between a black outline and a white background, are of course differences in intensity, and the visual system enhances such contrasts. It is this which is largely responsible for the ability to detect small differences in the visual field. The actual stimulus strength seems to be one aspect of a stimulus which is dispensed with quite soon in the course of visual information processing, the concentration being on form and pattern.

For visual images which are close to threshold, the size or extent of the stimulus has a critical effect on its perceptibility. It has been demonstrated that for targets with visual angles of less than 10 minutes, area has a significant effect on perceived intensity. In this case absolute threshold is a function of intensity times area ($T = A \times I$), and this is known as Ricco's law. It should be noted that on most maps, virtually all single line symbols will be smaller than this. As Gregory (1974: 19) points out, 'Smaller intensity differences may be distinguished with larger than with smaller areas.'

Symbols on maps which have to be discriminated on the basis of their relative size cannot be discussed on the basis of 'laws' that have been developed to reflect differences in intensity. In addition, the fact that so many map symbols comprise small forms and dimensions also has to be considered. The problem of discrimination in map use, therefore, has to take a variety of factors into account, and cannot be dealt with by taking only a single aspect of a stimulus into consideration.

Discrimination of dimension

Discrimination of dimension in a graphic image implies the ability to make judgements of the relative sizes of objects, and therefore the difference between them. In the case of maps this applies to the diameter or area of a point symbol, the width or gauge of a line, and the spaces between lines. The judgement of length or distance can also be regarded basically as a discrimination task.

The problem of size judgement has been thoroughly investigated by research psychologists, mainly by means of viewing sets or series of visual targets under artificially controlled conditions. Certain basic questions are immediately apparent. What is the smallest difference in size that can be perceived between objects which are only slightly different? Does this remain constant in discriminating beween groups of large and small objects? Does it vary between individuals, and to what extent is it affected by learning or practice?

Most of the experiments carried out by research psychologists with regard to visual images have been applied to judgements of length, the areas of symmetrical symbols, and the areas of irregular symbols. Although the visual comparisons of symbols representing volume have also been tested, it is difficult to distinguish these perceptually from area symbols as they have to be presented as two-dimensional figures.

With regard to map use, the problem can be divided into two parts. For most maps, differences in symbol dimensions are only used to indicate a qualitative difference, i.e. that as one line is thicker than another it represents a different feature or class. In some specialized maps, the task is much more difficult, as the differences in dimension are intended to be perceived as quantitative differences, and the map user is expected to judge by how much two or more symbols differ in size.

Qualitative discrimination

The critical factors when making comparisons between map symbols are the presence of many small symbols (small visual targets), and the effect of spatial separation. With relatively large symbols, a size difference of 10 per cent – for example, one line 10 mm in width and another line 11 mm in width – will normally pose no problem. But an equivalent difference between a line 0.1 mm in width and another line, 0.11 mm wide would be impossible to distinguish (even if the lines could be constructed to such measurements). Therefore the general principle that discrimination is reduced for targets with small visual angles is important, bearing in mind that measurements in tenths of a millimetre are common in map specifications. This is accepted empirically in map design, where differences of 50 per cent or more are common for fine lines or small point symbols.

In a multicolour map, contrast between symbols normally includes more than one variable at a time. For example, a black line representing a minor road will not only contrast in hue with a red line representing a major road, but is likely to be different in dimension as well. The problem of discrimination is generally more critical in monochrome maps, in which only contrasts in form and dimensions are possible for lines and small point symbols. As comparatively large symbols can only be used sparingly to indicate the most important features, it is inevitable that small differences in dimension must often be employed for the majority of point and line symbols. Generally speaking, it is easier to perceive differences in form than small differences in dimension. Whereas two fine lines of the same hue and similar gauges may be difficult to distinguish, a continuous line and an interrupted line are visually more distinct.

Quantitative discrimination

The most common case of quantitative judgement on maps occurs in the use of proportional symbols, that is, point or line symbols constructed to represent specific quantities. There is probably a more extensive literature on this topic than on any other aspect of discrimination as applied to maps. Indeed, it is interesting that an operation of rather limited significance in map use should so often be regarded as a major research problem!

The graphic elements of map symbols – form, dimension and colour – can be used to represent certain characteristics of the feature or phenomenon concerned. One aspect may be an indication of size, which may be based on an enumerated total (such as the total number of inhabitants of a town or area), or a measured quantity (such as the annual output from a mine). Broadly speaking, the function of a map dealing with this type of subject is to show the locations of the places concerned, and also their relative magnitude, by constructing symbols which are proportional to their total numbers or quantities.

It has been recognized for a long time that a map user confronted with such a map may be unable to judge the areas of the symbols 'correctly', that is, determine visually the quantities represented. To give a simple example, a population map may indicate the relative sizes of towns and cities by using circles proportional to their total populations, possibly ranging from towns with a few thousand inhabitants to those with several million. The circles are constructed so that their areas are equal to the population totals. The legend normally includes a series of symbols indicating fixed points in the total range, such as 100 000, 250 000, 500 000 and 1 million. The map author's intention is that a circle representing a population of 750 000 should be judged as lying midway between the 500 000 and 1 million symbols shown in the legend.

Although many tests have been carried out (see Brandes, 1976), with groups of subjects differing in age, sex and education, it always transpires that most size judgements of the areas of such symbols are comparatively poor; that there seems to be a general tendency to underestimate the quantities represented; and that the way in which the legend is constructed has a marked effect on performance. Although many tests have demonstrated that map users are unable to produce the desired judgements, this failure has generally been regarded as due to inadequate map design. The one obvious conclusion that can be drawn from so many investigations is that it is impossible for human observers to make such size judgements.

In the first place the human visual system has no means of 'measuring' the size of an object in a two-dimensional field. The judgement results from a comparison of the target symbol either with another symbol also present, or with a remembered image. In addition, exact size (the extent of the image on the retina) is relatively unimportant in human vision. The same object, seen at different distances, will produce retinal images of different extents, which is made use of in stereoscopic vision to perceive distance. Such objects in a natural scene are perceived as being constant in size, even though their retinal images vary with distance.

The comparison of a symbol on a map with one in the legend also raises questions of how the information is stored. If both are perceived simultaneously in central vision, small differences can be detected, but this does not mean that the amount of difference can be judged. If subjects are asked to make such judgements, the normal responses will be of the type 'about twice as big', or 'nearly half as big again'. As Baird (1970) points out, such judgements make use of non-dimensional scales. If there is a considerable separation between fixating the target image and then the comparison image, the differential threshold is increased, and two images which are only slightly different may be perceived as identical.

Attempts to explain and account for the underestimation of proportional symbols, particularly circles, are frequently detached from the realities of how people actually use maps. They also demonstrate the fundamental belief of the psychophysicist that if only the right 'law' could be discovered, it would be possible to predict the response to a given stimulus. In a classic work on size judgement, Baird (1970) reviews some 27 different experiments using proportional symbols. The results range from underestimation to overestimation. Each test concludes with a mean value for the ratio between actual size difference and perceived size difference. What is significant is that there are as many 'exponents' as there are tests, which clearly shows that both the type of test and instructions given to subjects are critical. Baird (1970: 63) concludes: 'At present, it is impossible to state that a particular exponent reliably reflects the psychophysical function for any given set of stimulus conditions.' It must also be remembered that these exponents are mean values, and therefore do not necessarily coincide with the performance of any indivdual person. Forgus and Melamed (1976: 50) make this point clearly: 'One of the major problems is the fact that there are sizable differences between individuals in the exponents obtained for virtually any stimulus dimension. In fact, even the correlation between exponents, when different response measures are used with the same stimuli dimensions, have been found to be very small.'

Baird goes on to show that the likelihood of poor judgement increases with the degree of difference between the standard symbol and the target symbol being compared. For most of these maps, 'standard' symbols are those appearing in the legend, with which the map symbols are compared. If the two differ greatly in size, judgement is correspondingly poor. Stevens (1975: 26) admits the basic problem: 'A series of experiments made it clear that if a standard is used, it should be placed near the middle of the range rather than at one end.'

One reason for underestimation of the areas of proportional circles can be deduced from interrogating any group of observers. Unless instructed otherwise, the intelligent map user will make use of the only visual clue present for which discriminatory ability is good, at least over a small range. That is, he will compare the two diameters, and adjust this mentally for area, making use of elementary geometry. Knowing that the area of a circle is greater than that indicated by the diameter, he adjusts the comparative difference upwards. Less motivated subjects, or those instructed not to calculate, will be more influenced by the comparative diameters, and therefore underestimate.

Teghtsoonian (1965: 398) observes that 'most of the replies indicated that the judgments were based on estimating a linear dimension of the stimulus figure and squaring it', and that 'veridical areal judgments depend on the availability of a linear dimension common to all the figures in the series'. The fact that such tests are also affected by 'instructions' reduces their relevance to real map-using tasks, in which a map user's responses are not being directed by an experimenter's instructions.

It would seem, therefore, that instead of contriving experiments to determine an absolute ratio between real size and apparent size, it would be just as effective if the proportional symbols were calculated according to the map maker's own visual judgement. This would be just as likely (or unlikely) to match the performance of any individual person as using a calculated exponent. The success of any particular solution will still be affected by the degree of separation between the symbols. This will vary considerably between a small book illustration and a large sheet map, but this aspect of the problem is generally ignored by the advocates of psychophysical testing.

If it is accepted that discrimination in human vision is comparatively good over a limited range, while adequate size judgement is not, it is not difficult to arrive at a practical solution to the use of proportional symbols. The size judgement task should be replaced by a straightforward discrimination task. Wright (1978) demonstrated this by constructing a proportional symbol map in which each symbol (circle) was clearly differentiated in size from its neighbours in the series by limiting the total number of different symbols, and by including a legend containing each size of symbol in order. All the map user has to do is to match a symbol against those in the legend which appear to be similar, and then to decide 'same' or 'different'. For the comparatively small maps used in his experiments, the size differences employed are perhaps greater than necessary. However, if the same system was adopted for a large sheet map, in which some of the symbols would be distant from the legend, the use of marked size differences would be appropriate.

It can of course be argued that the purpose of such a map should be to give an immediate visual impression, and not to invite the user to scrutinize the symbols in detail in order to judge their relative magnitudes. But users who are seriously studying a map are naturally inclined to identify symbols by reference to the legend. If this is not the intention, it would seem preferable to avoid the legend entirely, and limit the number of symbol sizes, as the map user's judgments are unlikely to be more specific than 'large', 'medium' and 'small'. Even so, a map user confronted with a set of symbols which clearly form the main subject matter of the map will naturally attempt to understand them, and to perceive what they represent. Because, in a general sense, small graphic images are a very poor device for representing differences in quantity, there is a strong case for limiting the employment of such maps quite rigorously.

Size scaling and quantitative discrimination

Quite apart from the correct identification of symbols, other map-using tasks involve the need to judge distance, direction and extent. In some operations, mechanical aids such as measuring scales, dividers and perimeter counters can be employed for actual measurements on the map. Nevertheless, judgements of approximate distance or length are often made, especially in using topographic maps, road maps and charts.

Gogel (1968) distinguishes between scalar perception and ratio perception (the latter also being referred to by other authors as size judgement or magnitude estimation). The difference is important in map use, because for some purposes a user may make an estimation of distance (length) in standard units, whereas for other purposes he may be expected to perceive differences in magnitude according to some scale given in the map legend. Gogel (1968: 126) states that scalar perception 'refers to some unit not simultaneously present. For the estimating human it is the memory of the ruler or meter stick.' On the other hand, ratio perception only requires the decision 'that one extent is larger, smaller, equal to or some multiple of another'. He also observes that precision decreases with either temporal or spatial separation.

Discrimination of length

Division of a line into two equal parts is a process which results in matching the two parts of the line until they are judged to be equal; in other words, until it is impossible to discriminate between them. Further subdivision can continue, but at each step the divisions are judged to be either 'same' or 'different'. Judgement of distance on a map can be more complex, but a common requirement is to judge the difference between the length of a known unit shown by either the scale bar or the grid, and to estimate the difference between this and the required distance. For example, many walkers and motorists must have visually 'counted' the number of inches or centimetres along a certain route to deduce the approximate distance. This also involves the short-term storage of the standard unit, as it is mentally compared with the distance required. The ability to store a unit length such as a centimetre or inch is considerably affected by practice. But it only operates effectively if the unit is small enough to be viewed in one fixation, and if the length being judged is relatively close to it. Although many map users could estimate a length of 5 cm with very little error, they would find it difficult to reach the same performance with a line of 73 cm. With relatively short distances, the whole length can be viewed within a few fixations and without head movements, with only a minute interval between them, and the fixations can be repeated rapidly. As the perception of small differences requires central vision, short-distance judgements can proceed more efficiently than those requiring many fixations and additional head movements. In those cases where lengths are estimated without reference to any visible standard unit, Hartley (1977: 622) reported that subjects in the experiment 'formed an image of the standard length'.

Unlike judgement of length, judgement of angle or bearing seems to have received little if any attention. However, it can be surmised that when necessary directions can be judged by approximating divisions between the main cardinal directions, and subdividing these distances into halves, quarters and so on, to arrive at the intermediate points. In this respect the procedure is similar to line subdivision, and involves the same basic procedure of matching two units to see if they are the same or different.

Discrimination of form

The term 'form' can only be applied directly to point or line symbols on maps; the apparently equivalent term 'shape' needs to be reserved for the spatial arrangement of geographical features. Therefore, form is an element in identification, whereas shape leads to recognition. Considerable research has been carried out into the perception of form, mainly with a view to understanding which forms are easiest to identify. Such investigations are usually conducted by asking an observer to detect a particular figure, such as a square or triangle, which is presented in a visual array containing a variety of forms. In fact the observer is being asked to match each of the figures against an 'image' of the target figure, which may have been verbally described (such as a square or a circle), or previously presented as an image. The comparison figures may be presented simultaneously, or in succession.

Such tests generally confirm what might be expected from the physiological evidence. Regular, simple, geometrical figures are easier to discriminate (i.e. are identified more quickly) than complex ones. Sutherland (1973: 167) reaches the conclusion that 'Symmetrical shapes are more accurately recalled by Man than non-symmetrical shapes of equivalent complexity, particularly if they are symmetrical about their vertical axis.'

The general view is that the pattern of light falling on the retina, representing the visual field, is not simply recorded and transmitted to the brain, but is analysed and processed throughout the visual system. This of course leads to the fundamental question of whether the brain stores a complete representation (or template) of each individual figure, object or shape, or whether each is broken down into a series of edges and angles. If the latter hypothesis is correct, then this would certainly help to explain that as regular figures are more easily processed, they can be identified more quickly.

Given that symmetrical and simple figures are more easily discriminated than complex and irregular ones, one important factor which applies to all map symbols is that discrimination is reduced with very small visual targets. For example, although a circle and a square, each with a diameter of 2 mm, can be easily distinguished, if they are sufficiently reduced in size it becomes difficult, if not impossible, to discriminate between them. Discrimination fails before detection. This may be further influenced by the degree of contrast they have with their backgrounds. Even so, if small point symbols are required (as they are for many minor features on maps), regular geometrical forms are more easily discriminated than complex forms. The 'pictorial' or iconic symbols often used for some features or activities on tourist or recreation maps, although often regarded as easier for the map user to understand, have to be comparatively large in size if they are to be discriminated with certainty.

In many maps the line symbols carry the major part of the total information. Lines, of course, can be used to represent linear features as well as outlines of areas. The form of a line may be continuous or interrupted. Although many line symbols are continuous, a great variety of line symbols can be devised, making use of both interrupted and combined forms. Because variations in single interrupted or 'pecked' lines can be difficult to discriminate with certainty, more complex combinations are often used to give greater contrast. The method can be exemplified from nautical charts, which often use lines interrupted by a number of dots, which can be easily identified.

Discrimination of form is also affected by separation. If the two symbols being compared are small and placed side by side they can be perceived in central vision in one fixation. If they are widely separated it may be necessary to carry out a larger eye movement between one fixation and the next, which results in a time lag in the comparison of the two images. The symbol first viewed must therefore be stored in short-term memory. If the second target is not found quickly, it may be necessary to 'refresh' the stored image with another look. This is particularly important with small variations in form. The total size of the visual field, and the spatial (and therefore temporal) separation between symbols, and between symbols and legend, is a factor generally ignored in psychological tests. If the contrasts in form between symbols are pronounced, errors in discrimination are unlikely to occur. But if the differences are small, requiring fine discrimination, the degree of separation in the visual field can be significant.

Colour attributes

The use of colour on maps is important, because it introduces a large number of variables which can enhance contrast, and therefore extend the number of perceptible differences which can be employed in discrimination. The effect is to aid legibility, and therefore to increase the total range of information which the map can present.

In colloquial speech, colour is generally used as a synonym for hue, as the spectral composition of a colour is its most obvious characteristic. It must be appreciated, however, that colour has three main attributes: hue, lightness (value) and saturation (chroma), and that the term 'colour' also embraces the differences which can be perceived in an achromatic image, black and white.

The visible spectrum extends from violet to red, approximately 400 nm to 700 nm, and within this band the principal spectral hues which can be distinguished are violet, blue, green, yellow, orange and red. The hue of a given colour can be reduced in intensity until it is imperceptible. In this case it becomes a neutral grey of the same value. This effect is known as desaturation. The lightness value of any colour must lie somewhere between white (maximum) and black (minimum). The various hues of the visible spectrum have different lightness values (see Fig. 23). At maximum saturation, they are highest in the middle of the spectrum, so that hues such as yellow and light green are the lightest. This is because the visual system is most sensitive to radiation at about 550 nm. Conversely, the hues at each end of the spectrum, violet and red, have the lowest value of the pure hues. It must be noted that many colours other than those which occur in the spectrum can be produced by combination (e.g. purple and brown).

In a printed map, the image is produced by superimposing printing inks on white paper. These inks cannot be 'diluted' or made more transparent in the printing press, and therefore their saturation is a function of their composition. Both a 'light' and a 'dark' blue can be printed separately, but neither will have the saturation of a theoretically pure blue. An artist can reduce the strength of a colour by adding white to it, or diluting it to make it more transparent, thereby increasing the degree of reflection from the white paper underneath. This produces a tint, which combines desaturation with an increase in lightness. Conversely, adding grey or black combines desaturation with a decrease in lightness, producing a shade. Variations in saturation and lightness are important in creating colour differences which do not depend entirely upon contrasts in hue.

In an image constructed on a white background, and seen by reflected light, the hues which have a low lightness value have the greatest contrast with white; those with a higher lightness value, such as yellow, have a much lower contrast with white. Consequently, if a yellow is further decreased by being desaturated, its contrast with white is rapidly diminished to the point where it will become imperceptible in small areas. Conversely, black, which theoretically does not reflect any light at all, has zero lightness value, and therefore maximum contrast with white.

If light waves are analysed into their spectral components, then any colour can be matched by a suitable combination of blue, green and red. If added together these produce white light, and therefore are called the additive primaries. If the light rays reflected from a coloured surface are analysed, then these will consist of white light which has been reflected through the filters of the pigments. They can be matched (in theory) by a proper combination of cyan, magenta and yellow. If added together then theoretically they produce black, and so they are known as the subtractive primaries. Cyan is white minus red, magenta is white minus yellow, and yellow is white minus blue. It is of interest to note at this point that although people generally regard yellow as a pure hue, just as much as green or blue, it does not appear among the additive primaries. Because printing inks are imperfect, it is normal practice to print a separate black. Because in each case any hue can be matched by a mixture of not more than three primaries, colour vision has generally been regarded as trichromatic. For a long period it was more or less taken for granted that the photosensitive cells in the retina must consist of cells (cones) which were responsive to blue, green and red. Physiological investigations have shown that although cones do differ in their sensitivity to different wavelengths, this only partly accounts for the complex processing which provides colour vision.

In multicolour map design, the contrast effect of different hues provides the most dominating visual clue in differentiating different symbols. Consequently, it is almost standard practice to use different hues for the main feature categories; for example, blue for water, brown for contours, and so on. For point symbols and fine lines, the hues chosen need to have sufficient contrast with white, and therefore it is normal to use the low-value hues such as dark blue, red and brown, and of course black, for the line images. Once again, the size of the retinal image has a marked effect on perception. Discrimination of different hues becomes more difficult with very fine lines and very small point symbols. A purple line, a dark blue line, and a dark brown line can be difficult to discriminate if they are very fine, even though the presence of a line can be detected. The common desire to use very fine lines to achieve an appearance of 'accuracy' and 'detail' is often opposed to the requirements of good discrimination. Because of the concentration of cones in the foveal and near-foveal region, maximum sensitivity to hue (wavelength) also requires central vision. Vision at the outer periphery of the retina, containing only rods, lacks any effective colour response.

Colour is used to maximum effect on a map when also employed for areal symbols. In particular, the recognition of geographical features is much enhanced when areas are differentiated by hue. Where colour is used extensively for areal symbols, then the combination of differences in hue, lightness and saturation, either as continuous surface colours or broken down into patterns, provides a large number of visual clues to identify different classes and subclasses. At the same time, complex colour arrangements may raise problems in discrimination, so that although multicolour maps enlarge the graphic possibilities, they also increase the probability of errors in the judgement of discrimination.

Discrimination of hue

To understand the way in which visual information processing works in the perception of colour, it is necessary to enlarge on the description of the visual system previously given. It has been pointed out already that the retina contains cells which are linked to the ganglion cells through receptive fields. The cones, which are sensitive to differences in wavelength, appear to respond maximally to wavelengths of about 440, 530 and 570 nm, which correspond roughly to the wavelengths which are sensed as blue, green and red. This of course assumes photopic vision.

According to Ditchburn (1973), colour discrimination at the retinal level is controlled through three sets of signals: the difference between white and black; the difference between the amount of red and the amount of green; and the difference between the amount of yellow and the amount of blue. This is not proposed as an alternative to the trichromatic theory in the optical sense, but as an explanation of the way in which the information is processed (Fig. 20). The basic principle of lateral inhibition has already been described (see Fig. 16). In this instance, the excitatory and inhibitory effects are a function of wavelength, and therefore spectrally opposed, instead of being spatially opposed. It seems that some cells respond to 'red' wavelengths by an excitation but are inhibited by 'green', and correspondingly some cells are excited by 'yellow' but inhibited by 'blue'. The presence of such cells with paired responses has been detected in the retina, the lateral geniculate nucleus, and the visual cortex. These react with paired responses, red–green being one pair, and yellow–blue the other (Fig. 21). It is of interest that as long ago as 1878 Hering suggested an opponent-colour theory of this type. It has also been known for a long time that coloured after-images are paired (green after red, blue after yellow), and this also suggested paired responses. The operation of the opponent-colour theory was established by the work of Hurvich and Jameson (1957, 1974). The analysis of signals in the retina is completed by cones which respond to all wavelength signals, and the function of which seems to be to record luminosity in terms of intervals between white and black.

Wyszecki and Stiles (1982) suggest that both the trichromatic theory and the opponent-colour theory can be merged into what they call zone theory. In this, it is suggested that colour vision is initiated by the

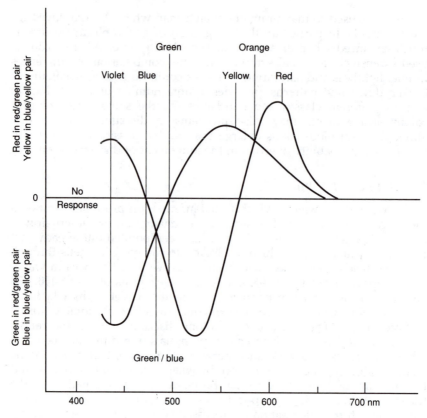

Figure 20. Opponent-pair theory of colour perception

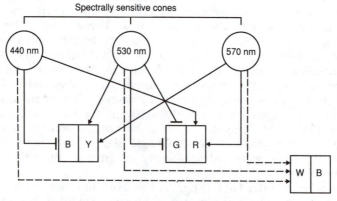

Figure 21. Opponent-pair responses to trichromatic input

trichromatic signals, but that at a second stage the neural network creates three new signals, one achromatic and the two opposed colour pairs.

The actual processing of the signals from the retinal cells appears to take place primarily in the lateral geniculate nucleus. Although the analysis of signals from retinal cells is of almost bewildering complexity, the opponent-colour theory does have the advantage of giving yellow a specific function in hue discrimination, which accords with the sensation of colour normally experienced.

Simultaneous colour contrast

The process of spectral inhibition has also been invoked to explain the phenomenon of simultaneous colour contrast. For example, if a red field surrounds a neutral grey area, the grey will tend to appear very slightly green. Hurvich and Jameson (1974) explain this by proposing that neurons excited by the long wavelengths (red) will send inhibitory signals to adjacent areas (such as the grey), and that these inhibitory signals will produce the opposite response of green. In this manner, both spectral and lateral inhibition may operate together.

Complex colour combinations for areal symbols on maps, such as lines of one hue over a background of another, may involve such an inhibitory process. Although they have not been analysed or investigated, such slight shifts of perceived hue may affect discrimination. For example, patterns of green lines can sometimes appear slightly more 'blue' than anticipated, possibly because the red component of the intermediate white spaces is being laterally inhibited by the green.

Discrimination of lightness and saturation

The problem of judging differences between stimuli which differ only slightly has been referred to in relation to the comparison of lengths and areas. Similar considerations apply to variations in lightness and saturation within any one hue. Many systems of colour analysis exist, in which it is necessary to indicate not only differences in hue, but also in lightness (value) and saturation (chroma). In one of the best-known colour description systems, the Munsell System, a three-dimensional colour solid or diagram is used to represent the three properties. Hues are arranged radially; value (lightness) extends vertically from black (zero) to white (maximum) at the top; and chroma (saturation) extends horizontally (Fig. 22). The different steps in all three scales are derived from judgements of least or just noticeable differences, as recorded by a group of observers for a specific visual field.

From the point of view of discriminating colours on maps, three factors are important. In the first place, for hues which have a high lightness value (such as yellow) only a few differences in saturation are apparent. Conversely, for a dark blue or red, a larger number of perceptible differences can be made between maximum saturation and a neutral grey of the same value. Second, the perceptible differences in

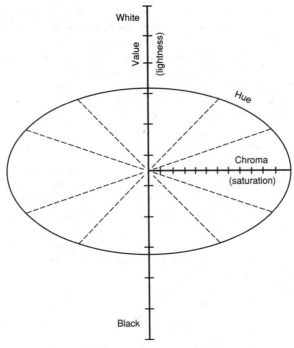

Figure 22. Basic arrangement of the Munsell System: hue, value (lightness), and chroma (saturation)

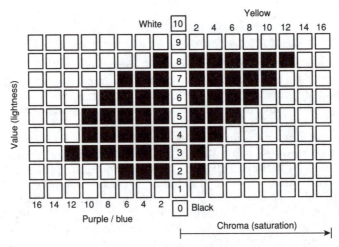

Figure 23. Section through the yellow–purple plane of the Munsell System

value (lightness) and chroma (saturation) are not symmetrical for different hues. Whereas yellow at maximum saturation has a lightness value on the Munsell scale of 8, the purple-blue at maximum saturation has a lightness value of only 3 (Fig. 23). In the third place, the ability to discriminate between different saturations of the same hue is strongly affected by the area of the image, and by spatial separation. If two greens of slightly different saturation are placed adjacent in the legend, the difference between will be more apparent than if two small areas of the same greens have to be compared when widely separated on the map.

Simultaneous comparison in central vision is more acute than when images are spatially separated. In addition, the presence of two green symbols in the legend implies that they are intended to be different – and human perception is always affected by the tendency to see what is expected. Where fine discriminations between colours which vary in all three attributes are concerned, performance is considerably affected by practice.

Fine discriminations in saturation and lightness are only possible if there are no other colour distractions. If other colour elements, such as lines crossing the areas, are introduced, then the ability to distinguish slight differences in saturation or lightness is decreased.

Because the design specification of map symbols is often begun by constructing a prototype legend, in which the symbols are considered in isolation, the effect of area, shape and spatial separation is not always anticipated. Particular combinations of colour differences on maps can therefore pose problems for the map user. For some types of specialized maps which have a 'standard' series of colour area symbols (such as some geological maps), the context of a particular map (with very large areas of one colour symbol and very small areas of another) can present discrimination difficulties. Sometimes this is overcome by adding an additional clue (such as a numerical reference) to the individual areas, to make identification certain.

The use of colour charts as a design aid, in which series of variations and combinations of hue, lightness and saturation are systematically arranged, can also lead to the use of colour variations which are too close to threshold. There is at least one geomorphological map which shows five different tints of yellow in the legend, of which only two can be identified with any certainty on the map itself.

Identification

If a map contains only symbols which can be identified in the legend (or in a legend separately available) then the problem of identification would seem to be straightforward, in the sense that it should depend on discrimination in conjunction with short-term memory. Repeated reference to the legend symbols should lead to their establishment in long-term memory. This of course presumes that what the symbol represents is also understood by the map user.

Different types of map use, however, will require different levels or types of identification. Ellis (1972: 161) points out that 'the task of identification requires a unique response to each stimulus.' But Fellows (1968) goes beyond this to make a distinction between two sorts of identification: identification and match-to-sample. If a single target is sought on the map, for example the symbol representing a canal, then other map symbols will be matched against it individually, the response being 'same' or 'different'. But if an area of map is being studied at a higher level of interpretation, then all the symbols must be correctly identified as they are encountered. In this case the required map symbols must be held in memory and made available for comparison as they occur. The question of whether this elimination task is carried out serially or in parallel has been argued at length (see Vickers, 1979).

Several researchers have come across evidence that seems to suggest that in match-to-sample the identification of images similar to the one required differs in some way from the rejection of dissimilar ones. Solman (1975: 411) reports that Neisser was one of the first to observe that 'the perception of irrelevant or field items differed in some fundamental way from the perception of target items. The irrelevant items were processed in some superficial or elementary way to select items with target-line features or characteristics.' Keuss (1977: 371) also suggests 'a dual process model that is based on an identity reporter for the faster "same" response and a difference detector for the slower "different" response.' On the other hand, Gould and Dill (1969: 317) report that 'subjects looked longer at target patterns than at non-target patterns.' This suggests that the features in a fixated 'pattern' are compared with features in a memorized standard 'pattern' in a serial or sequential manner, and that this comparison terminates upon detection of a critical difference between the two.

The different conclusions reached by these researchers are likely to reflect the different degrees of similarity or dissimilarity. If the map user is searching for a single target, such as a red road, then the reaction 'same' will be almost instantaneous, and very different symbols, such as blue rivers or brown contours, will be immediately dismissed. But if she is seeking a complex geological symbol, which is itself similar in characteristics to several other symbols, then each occurrence of this target-like group will have to be scrutinized at greater length before the decision is reached.

Identification and geographical context

Although the operation of identification can be discussed in terms of the symbols given as 'examples' in the legend, these of course are out of their geographical context. In actual map use, context has a powerful effect. The characteristic forms of geographical features are also 'familiar', although the degree of familiarity will also be a function of experience in map use and geographical knowledge. The contrast between a road and a river is not limited to the employment of different map symbols; it is also a function of the degree of regularity and sinuosity anticipated or expected. In this respect, typical linear arrangements can also be 'recognized'. Although it has been argued that it is useful to differentiate between identification and recognition in relation to map use, it is clear that a recognition factor can also operate in identification, due to the known (and therefore familiar) character-istics of specific features.

The reader may have noticed that the three map examples introduced at the beginning of this book do not include legends. Strictly speaking, therefore, identification is impossible. Yet the maps are intelligible because the characteristics of the limited number of features represented (coastlines, contours, roads, etc.) provide their own clues to identification. The point is well made by Imhof (1951: 100, Fig. 194).

If identification has to be assumed, then presumably it can only take place through a process of interpretation. Once again, it underlines the fact that in actual map use it is difficult to draw any exact line between 'processes' such as identification, recognition and interpretation, or to regard them as a fixed sequence of events which must occur in a particular order. It is possible that interpretation, based on characteristics and expectations, can lead back to identification as a process of verification.

Identification and perception

In dealing with the basic problems of detection, discrimination and identification, it has been suggested that 'map reading' at its most elementary level could not be described solely by these processes. The identification of a brown line by reference to the legend may lead to 'contour', but the meaning of the term has also to be understood. If 'perception' includes comprehension, then map reading can only place when the meaning of the symbols is understood.

In actual map use, at least with general maps, the need to understand the 'meaning' of the symbols is so obvious that it is unlikely to be considered at all by many map users. Yet the point is important in considering what the term 'perception' refers to. Is it a cognitive process, which depends on knowledge, or does it simply describe the response of the visual system to an external stimulus? There is no agreement on this point, and indeed many writers refuse to give perception a specific definition. Yet psychological tests of discrimination and identification, making use of arrays of meaningless 'figures', do not involve comprehension in the sense usually implied in reference to map reading.

Presumably a person can 'read' a completely unknown language by correctly identifying the letters which form the words, but this is hardly what most people would regard as 'reading'.

Once the meaning of a symbol is understood, interpretation is possible, and interpretation always involves the user's knowledge, not simply visual reactions. The processes of detection, discrimination and identification must be carried out, but by themselves they do not permit the user to extract any information. At the most, they are only a preliminary phase of map use, and they have to be involved in, or followed by, interpretation.

Identification and memory

As immediate or short-term memory can only retain a limited number of items simultaneously, it is clear that scanning an area of map must involve selective attention. A whole area of a map cannot be 'stored' at one time, but must be examined in a series of fixations to extract significant or pertinent features. This raises the question of how much material can be dealt with, and in particular the amount of material that can be acquired through one fixation. In experimental tests, the difficulty is to ensure that a single fixation is isolated. Therefore, in some experiments subjects are restricted to a very brief display of a number of items in a visual field, the object being to discover how many can be perceived and remembered. According to Sperling (1960), subjects are usually able to report up to six items from brief stimulus exposures. This limitation of short-term memory may have some bearing on the density of 'detail' in a map which a map user can comfortably or conveniently handle. In general, people are reluctant to operate close to the threshold of performance, and find it fatiguing, as concentration is difficult to sustain. If too many objects are present within the extent of a single fixation, then the rapid identification of these may overload the short-term memory capacity, leading to repeated fixations of the same area, and the reaction that the map is 'difficult to read'.

The limited capacity of short-term memory is also referred to as memory span. It is clearly important for any conscious person to be able to 'keep things in mind', whether or not these things consist of stimuli received externally, or are the products of internal memory or thought. Short-term memory is also described as active memory, as the normal process is to retrieve items from long-term storage and bring them into an active state. In any type of map use which is time-constrained, the limitations of memory span may be critically important. Tests of aircraft pilots and navigators using charts for visual flight control generally report a desire for a reduction in the total information presented and the removal of 'unnecessary' items. This can be construed at least in part as a desire to lessen the amount of information which has to be briefly retained and processed.

The means by which information is added to long-term memory is not fully understood, although it is a problem which concerns educationalists in particular. There is little doubt that rehearsal is important in all forms of learning. But for most people, whereas some things seem to be entered with little difficulty, others prove virtually impossible. The general view is that the chief difficulty is not storage but retrieval. As Herriot (1974: 7) puts it, 'there is more material available at the time of recall than is actually accessible (i.e. can actually be retrieved).' It is also one aspect of the brain's operation which seems to differ markedly between different people. How much of this ability is innate, and how much the result of practice and learning, continues to be debated.

Memory and map use

In the three examples of map use in Chapter 1, the requirements for memory are quite different. In the first example (see Fig. 1), the task is primarily match-to-sample. In this case the map user is looking for a name, and his ability to identify correctly a block of letters forming a word reflects the great deal of practice he has had in formal education in this particular visual task. In the second example (see Fig. 2), short-term memory may limit the number of road junctions or changes of direction that can be recalled. In this case there is a time constraint because failure to recall correctly will be demonstrated as the journey progresses. Verification will be a check that the intended sequence has been memorized correctly. Some of the information about roads in the area may pass into long-term memory, especially if the journey is repeated, but this is not necessarily part of the intention.

In the third example (see Fig. 3) long-term storage is likely to be the aim. This may be accomplished through a rehearsal and self-testing procedure, by studying the map, leaving it for a while, and then checking whether the remembered 'facts' agree with those presented. Long-term memory will be able to retain a large number of items, but these are more likely to be propositions or conclusions than exact visual images. A subsequent attempt to use the information will also depend on the ability to recall particular items, and bring them into active memory. These general differences also involve the degree or duration of attention which is given to the map-using task. If a map interpretation is sustained, then input from short-term memory and withdrawal from long-term storage will interact over a period. Maintaining such a level of concentration is arduous, and again, it is a process which benefits from training.

Memory and recognition

The ability to perceive objects as being familiar has already been given a particular definition in map use, mainly to distinguish it from identification of symbols. Recognition in this sense is applied to the perception of geographical features, or even the patterns typical of some phenomena. At the basic level it requires only the judgement that a perceived feature is the same as one previously experienced, and it may or may not be accompanied by the attachment of a geographical place name.

In map use, two types of recognition can be discerned. The first, and more general one, is the immediate recognition of a shape – the outline of a country, continent, lake, etc. In most cases such images are 'small-scale', and will be the product of learning from maps, as in many cases they cannot be perceived in their entirety directly. On the other hand, many orientation tasks depend on the ability to recognize a topographic feature or group of features, either by direct comparison with the map, or from memorized information. In the second map example (see Fig. 2), a particular road junction may be recognized by road alignment, or the presence of landmark features. Recognition has to be applied to this procedure, as the characteristics concerned will be unique to that particular locality. The process of anticipation in following a route often depends on the ability to detect such a congruence of features, and realize that they signify a particular location.

It is clear from this that memory cannot be regarded as simply containing a number of discrete images which can be brought into action by a stimulus identical to the initial one. Memory must be capable of processing, adapting new stimuli which may be different in detail to previous experiences. The need for this is made clear by many investigations. Young (1978: 79) states that 'Memories are thus physical systems in brains, whose organisation and activities constitute records or representations of the outside world, not in a passive sense of pictures but as action systems.' Young (1978: 90) also states that 'Memories would be of little use if they only allowed the establishment of a connection between a few receptors and the nerve cells that activate the appropriate muscles . . . . Any effective neural memory system must have a mechanism for generalisation.'

Transformation and visual information processing

As a map is not a 'picture' which corresponds directly with the appearance of the terrain, or any other phenomenon, as a single image, orientation must involve an act of interpretation capable of converting the map symbols into the individual features they represent, transforming the orientation as required, and changing the scale. It is in fact a good example of higher-order processing. It is clear that if the perception of a map only involved the storage of an 'impression' of the map, rather than a constructive analysis of it, the transformation processes referred to would be impossible. Evidently the brain is capable of analysing the incoming information and subsequently carrying out different arrangements and combinations. Even if a very simple 'map' could be stored completely – and whether this is possible is arguable – it would not be much use unless the information contained in it could be operated on in some way.

If the basic processes of transformation of scale and perspective, interpretation from two to three dimensions, and recognition of objects at different orientations, are examined further, it is of interest that the first two seem to have received very little attention in psychological literature or cognitive studies. Recognition of objects or pictures has been investigated in detail, and it is well established that two-dimensional representations of three-dimensional objects, rotated in the same place, or in depth, can be recognized, although the process is a slow and deliberate one. Shepard and Metzler (1971) demonstrate this with a set of line figures, rotated in the same plane, and in depth.

Transformation of scale (from map to reality) may be comparatively straightforward when both the map and the terrain within the field of view are compared simultaneously. 'Real' dimensions can be perceived and matched against the relevant distances on the map. The chief difficulty lies in the different degree of 'generalization' of near and distant objects in the visual scene. The perception of distance through stereoscopic vision is, however, a normal visual function, and therefore the judgement of distance in relation to apparent size is reasonably well developed.

Transformation from the orthogonal projection of the map to the central perspective view of the terrain is also comparatively simple when both map and terrain are viewed together, for the limiting effect of a low angle of view on the ground is counterbalanced by the more extensive representation of the map. Obviously one of the advantages of the map is that it overcomes the limitations of viewing the landscape from a single point, from which many lines of sight may be obscured.

It is when a map is used in anticipation of views which cannot be perceived or before they are encountered that more difficult problems emerge. In this case the interpretation of the terrain has to be derived directly from the map, and 'memorized' if it is to be used to 'recognize' locations or features when they are actually encountered. It is in this process that a purely visual transformation is unlikely, that is, an attempt to construct an 'image' directly from the map. What is more likely is the deduction of a limited number of conclusions about key points, the relative position of landmarks, and so on, which will provide a framework of clues. In this case it is also difficult to see how scale change takes place. As the neuronal circuits in the brain which actually 'store' the information cannot be said to have any scale, it seems more probable that features on the map will be recorded in relation to their relative positions, and that the true 'scale' will only become apparent when the scene is perceived in reality. The sheer size of a major mountain range can be surprising if it has only been 'perceived' previously from a map representation.

When the map deals with an area which cannot possibly be perceived directly, such as a continent or large country, a different set of factors must apply. In the abstraction task of the third example (see Fig. 3), the scale of representation is relevant not to any perceived reality, but to the degree of detail needed to present the subject matter. As the 'true' or real size of a continent cannot be visually imagined, concepts of large areas are more of a schematic framework for locations than a picture which can be recognized. Although the term 'mental map' is often applied to such internal representations, they are quite unlike maps because they are personal, fragmentary, incomplete and presumably frequently erroneous. Indeed, one of the main reasons for making maps must be that 'mental maps' are inadequate as useful stores of locational information. It is by no means clear as to whether our apparently limited ability to memorize any great quantity of locational information is due to some innate limitation, or because the learning (and memorizing) of spatial information is rarely regarded as a serious or valuable activity.

Visualization

Although visualization, the construction of a mental image or representation without the stimulus of direct perception, is often referred to, it is by no means clear that it results in a complete and perfect 'picture' which can be held in the 'mind's eye'. The process of matching features interpreted on the map, and the terrain – such as the course of a river, the relative positions of buildings, the alignment of the edge of a wood – does not necessarily have to take place as a single action, nor does the contemplation of a map inevitably lead to some visual representation. Indeed, it is more likely that key features or points are extracted and their relative positions established, and that these may be recorded as a set of verbal statements rather than as 'imaginary' images. It is also unlikely that more than a very small area of a map could be interpreted simultaneously and 'visualized' in its three-dimensional form. In some respects this is a particular case of a general problem: whether spatial information is stored in analogue form (as a series of pictures which can be recalled), or whether it is broken down into a set of critical features, which can be further processed to extract different propositions that can be dealt with verbally.

It should be noted that the term 'visualization' is now extensively used to refer to the creation of an image on a screen by constructing a model (usually representing three dimensions) derived from the data being examined. This, of course, is intended to aid visualization and interpretation (see Hearnshaw and Unwin, 1994). Although regarded as potentially useful in many kinds of analysis, it is fundamentally different from visualization as an act of perception and visual imagination.

Visual and verbal processing

The question of the relative roles of visual and verbal processing has been widely discussed and investigated. As most visual processing in learning consists of reading from printed texts, this has naturally been the main area of research. Some attention has been given to the recognition of pictures. It is generally accepted that briefly viewed pictures of objects and scenes can be remembered, even though they cannot be named, and that therefore there must be some means by which visually perceived material can be recorded and stored without any verbal mediation. It has also been observed many times that memories of objects, places, or even abstractions can be assisted by association with visual images.

On the other hand, other investigators take the view that verbal processing is essential to thought, and therefore to memory. Norman and Rumelhart (1975: 17) state that 'The fact that a person "perceives images" when recalling perceptual experiences from memory does not mean that information is stored within memory in that way. It only implies that they are processed as images.' Pylyshyn (1973) argues that pictures cannot be stored as complete point-to-point representations, because both the processing and the storage for each would be enormous and therefore overload both the sensory system and the brain.

Given the extensive interconnections within the brain, the most likely answer is probably that expressed by Baddeley (1976: 234). He describes long-term memory as 'a single semantic abstract memory system which contains both linguistic and pictorial information and which can be accessed equally well by words or pictures'. A similar point of view is expressed by Paivio (1971: 508) in his reference to 'a two-process theory in which imaginal and verbal symbolic processes are functionally coordinated to concrete and abstract task performance.' In dealing with the relationship of propositional information and imaginal representation (in terms of mental storage), Norman and Rumelhart (1975: 17) comment that 'As soon as a person realizes that mental images – even analog ones – are not the same as photographic reproductions of the original sensory experience, then the door is open for the merger of the two apparently conflicting ideas about representation.'

In an interesting experiment on a somewhat unlikely example (an attempt at complete storage of the information in a map). Thorndyke and Stasz (1980: 173) concluded that 'In both experiments individual differences in recall of verbal information and in the use of verbal learning were much smaller than for spatial information. Further, virtually every subject learned more verbal than spatial information from the maps.'

In an educational environment which is dominantly linguistic, it is possible that people depend more on verbal processing because they are more competent at it through practice. On the other hand, it may be that innate differences in verbal and spatial processes cover different ranges of ability. The problem illustrates the difficulty of analysing activities in the brain on the basis of conscious behaviour.

In the case of maps, it is likely that although people frequently make statements which are essentially deductions from maps expressed in words, they can also imagine at least the main forms of familiar terrain, or even part of a map itself. Such visual images would seem difficult to maintain or repeat exactly, and consist more of an ordering of things, rather than something resembling a photograph.

Visual search

The expression 'visual search' is widely used in psychological literature, and it is often referred to as an important function in the use of maps. However, it covers a variety of visual processes, and has to be considered with some care.

In psychological research it is used mainly in two contexts, which can be identified briefly as subconscious and conscious operations. When an object is perceived and recognized, it is assumed that features of the object have been extracted and processed, and in this sense the object itself has been 'searched'. The point has already been dealt with, and it is evidence that this type of 'search' is carried out at a subconscious level, though a subject may be able to describe his behaviour during the process. The relation between feature detection and pattern recognition (to use the psychologists' terms) has been studied in detail and Coltheart (1972) makes the point that much of the capacity to discriminate different forms appears to be innate.

It is assumed that the ability to detect and analyse features is fundamental to the visual search task, in which, for example, a subject is confronted with a display of different objects in a visual field, and asked to search the display to locate a particular object or target. In this case the subject is consciously attempting to perform a task, and psychologists are interested in finding out how the characteristics of different objects (form, size, colour) or arrangements affect the efficiency of the searching process. This assessment is normally based on the time taken for the subject to locate the target or targets. Such research tends to do little more than confirm what has been known empirically for a long time. The fact that 'Colour-coding' is a powerful element in differentiating targets, even if they have the same form and dimension, is not exactly unfamiliar to map makers!

On the other hand, a picture or an image can be scanned in order to appreciate or understand it. In this case there is no longer any single target, the location of which will terminate the search, and the visual search is likely to consist of a series of fixations which sometimes jump from one point to another. The sequence of such fixations has been recorded in several investigations, and these experiments show that fixations are directed to significant points, usually sharp contrasts and edges (Gaarder, 1975; Castner and Lywood, 1978). Young (1978: 119) makes the point that in normal perception, 'Vision is a dynamic process, using a series of scans, but these are not rigidly determined as in a television raster. They are varied according to the nature of the scene itself and the previous experience of the individual.'

It is clear that both types of visual search have a parallel in 'looking for' things on a map. It must also operate in 'looking at' a map, and therefore it is necessary to consider different levels and types of search in map use.

In the first example (see Fig. 1), the simple locational task involved only a limited degree of search. The indexing system of the atlas is devised to restrict this search to a relatively small area, in which the map user knows in advance that the object can be found. As only one name is required, the non-target items are easily dismissed. In the second example (see Fig. 2), the map user is also 'searching' but only within a limited area of interest, concentrating on the available roads and those pertinent to the desired route, creating a hypothesis for a likely route by selecting certain roads, and checking this against alternatives. As the minor roads have the same symbol, and therefore the same visual character, the search is directed at their spatial arrangement and interconnections. There is no single 'answer' to the question of which route to follow, and therefore the termination of the search depends not on finding something, but on judgement.

In the third example (see Fig. 3), the map user also searches the map, but neither to locate a single feature nor to resolve a problem The objective is to comprehend the total spatial arrangement, and to incorporate this pattern, or at least elements of it, in long-term memory. The spatial arrangement will be easier to understand, and make more sense, if it can be located against a framework of the real world previously familiar to the map user, or actively constructed during map use. Such a framework is likely to consist of a very simplified topographic 'outline'.

This type of search is much closer to the process by which people 'search' a picture or photograph, the pattern of eye movements being at a subconscious level. In research directed towards analysing how people look at pictures, there is a fair amount of evidence that people concentrate on edges, sharp discontinuities, and contrasts – which is what would be expected from a knowledge of the physiological processes of vision. But the particular series of fixations seems to be unique to each individual, so that although the eventual result may be similar, the path to it is different.

Looking at pictures does not necessarily have any 'goal' apart from a pleasurable sensation. People who like maps can also regard them in this way – hence the use of maps as wall decorations. But the searching involved in looking at a map may be finite, or a matter of judgement, or open-ended.

There are other examples of map use in which different levels of search are involved. A person looking for a suitable camping site, or a likely place to fish, is 'searching' not for some specific item of map information, but for the congruence of certain features which will match her knowledge or her preconceptions. Searching the map may even offer possibilities which previously she had not anticipated. She will not necessarily find what she is looking for, but even this is 'information' of value. This kind of search involves highly sophisticated interpretation which balances what can be perceived on the map with existing knowledge and intentions. It is not a random search in a meaningless array, for the topographic features represented on the map have their own physical relationships. Such a search is likely to be a combination of verbal reasoning and visual imagination, and it underlines again the critical importance of interpretation in map use.

Maps and visual perception

Psychological studies of the perception of form and pattern pay a great deal of attention to the recognition of shapes and objects. It must be realized that the ideas so developed, in particular those of the school of perceptual psychology known as 'Gestalt', are concerned primarily with how people recognize objects by direct perception of the external world. Frequently the visual data are incomplete and irregular, and appear at different distances and in different orientations. As Oatley (1978) points out, real objects often have to be constructed by distinguishing surfaces from ambiguous sets of edges.

The general point of view of Gestalt psychology is that the ability to detect and recognize shapes involves separation of figure from ground, continuity through interrupted lines, and so on, and that these processes are innate and independent of learning. Although many of these tendencies are observably correct, the notion that such theories provide a basis for an explanation of perception is firmly disputed by many other psychologists. Hamlyn (1957) presents the arguments against quite convincingly. Fundamentally the gestalt ideas are behaviourist, in that the assumption is still made that given stimuli will result in a particular response, because of the innate tendency of the organism to react in certain ways. The gestalt view emphasizes that figures or shapes are seen as 'wholes', which seems to be at variance with the physiological evidence, which suggests that forms are composed and stored by analysing them into discrete units.

Theories devoted to the direct perception of the external world, with the problems of perception of depth and distance, variations in orientation and illumination, and so on, are not directly relevant to the perception of a symbolized image, such as a map. In real life a person learns to classify certain objects as 'tree', even though those encountered close at hand may bear little resemblance to dark shapes seen in the distance. The individual trees are unique, and the response 'tree' results from a classificatory process. On a map, the classificatory concepts are stated, even if only in outline, and the symbols defined. The map does not include many different representations of the same thing, and then expect the user to classify them correctly.

The distinction between direct perception and the perception of a map is clearly illustrated by the photomap. A photomap combines certain aspects of the map with a transformed photographic image of part of the Earth's surface. The patterns of light and dark in the photograph (its tone and texture) have to be interpreted by the user, so that features in the terrain can be correctly identified, or at times recognized. Where the photographic representation – which is a picture – of the real world is too ambiguous, or insufficiently clear, so that all the features in a class cannot be identified with certainty, they may be replaced by a cartographic symbol, or an additional visual clue may be added, such as showing them in a particular colour.

Both the making and using of maps involve visual perception. The cartographic task is to devise the map so that the user's visual perception – to whatever end it is directed – is not impeded by problems of detection and discrimination. Therefore the cartographer has to make decisions depending upon his own visual judgement, and his anticipation of the user's discriminatory abilities.

These issues lie at the heart of cartographic practice. As a map is an artificial construction, 'legibility' is deliberately sought by using the graphic variables of form, dimension and colour to introduce sufficient contrast to permit detection and discrimination. In terms of some of the basic visual processes referred to previously, it is clear that in general minimum symbol dimensions are kept well above detection threshold, and the contrast between symbols is normally well above discrimination threshold. As the 'figures' (i.e. shapes in plan) are continuously outlined and/or represented by a change in surface colour, any problem of distinguishing figure from ground ought to be, and can be, anticipated in map design. The most common examples of 'bad' maps are either those in which the map user is pushed towards the thresholds of visual discrimination, or those which demand a discriminatory ability beyond the capacity of the graphic image to support. For the latter, maps using 'proportional symbols' are the most frequent offenders. It is quite remarkable that the prolonged investigations into proportional circles in particular have not yet reached the obvious conclusion: that the exact representation of quantities – however desirable it may seem to a map author – is contrary to the evidence of visual perception, and that there is no magic solution through some cartographic 'technique' or the manipulation of graphic design.

Perception and interpretation

This brief study of some of the major factors in visual perception and their relevance to understanding maps necessarily breaks down perception into a series of processes or stages, which are linked into a sequence. The study of perception, by both physiologists and psychologists, has sometimes to concentrate on particular mechanisms or operations, such as the reaction of retinal or cortical cells to stimulation, the capacity of short-term memory, and so on. Research of this type gradually builds up greater knowledge, and the understanding of how visual perception 'works' steadily increases.

Given the remarkable scientific discoveries of recent years, it might seem that in a relatively short time a complete description of visual perception could be attained. Yet understanding how we 'know' things seems to be as elusive as ever. The principal argument is whether perception – in the sense of knowing about something through a visual process – is simply the end product of a series of processes beginning with a stimulus, or whether it must involve some interpretative act initiated by the organism. Is knowledge supplied through visual processing, or is it constructed? Many of the scientists who are concerned with the larger view of how the brain works remain dissatisfied with theories which suggest that describing a set of particular processes will lead to a satisfactory explanation of how the mind operates on data supplied by the senses. As Hamlyn (1957: 34) puts it, 'To say of someone that he has a sensation does not ordinarily commit one to the view that he has perceived anything at all.' The same opinion is expressed by Bellin (1971: 93): 'sensation in itself is not considered to provide direct knowledge.' The point is argued in detail by Norman and Rumelhart (1975: 18):

We know that the perceptual information undergoes considerable transformation as it is processed by the sensory nervous system. Physiological mechanisms pull apart the components of the arriving signals and perform intricate types of frequency, temporal and spatial transforms on the signals. Feature detectors pull out significant parts of the wave forms. And then context, meaning, and past experience play a large role in the interpretation of the information, evidently in the very initial stage of perception and pattern recognition.

Partly because of the ambiguity of the term 'perception', cognition is used more and more to refer to the psychological study of human knowledge acquisition. Cognitive studies generally emphasize internal mental factors, such as memory, learning and attitude, rather than just input through the senses (see Anderson, 1990). Yet it is clear that the mind operates as a single unit, and cannot be said to distinguish between perceptual and cognitive operations.

These questions are important in map use, because they help to clarify the issues of map interpretation and its investigation. If 'map reading' involves no more than responses to controlled stimuli, then psychophysical studies should lead to an improved understanding of the map user's problems. But if map use is a cognitive act, designed to acquire and integrate new knowledge on the basis of the map user's needs, interests and existing knowledge, then progress is more likely to be achieved by studying and observing people engaged in real map-using tasks. The analysis of factors in map use in this respect is a particular case of the general problem referred to by Taylor (1964: 270): 'In fact the question we have been discussing is not whether a science of behaviour is possible, but rather, which direction should we proceed in order to develop such a science.'

If the behaviourist or phenomenological approach is accepted, then it should be possible in the long run to produce a scientific 'model' of visual perception as a series of processes. If the cognitive view is taken, then both the complexities and the mechanisms of the mind may turn out to be very different to those suggested by 'scientific' models. Indeed, an effective analogy for the working of the brain is presently lacking. As Blakemore (1977: 103) points out:

Description by analogy seems less successful for the nervous system than for other organs in the body. The heart is like a pump in its action and the kidney is like a filter because they are a pump and filter. But however similar in its properties mental memory is to a computer core or a hologram, the brain is not a set of magnetised rings or a laser-illuminated photographic plate.

And he concludes:

Unfortunately, the physical structure and mechanisms of operation in the brain are so unlike those of any piece of machinery made by man that analogies are usually weak.

PART TWO

The map as a symbolic representation

CHAPTER 4

Signs and symbols

In dealing with the problem of visual perception it has been shown that the ability to make sense of a map depends in the first instance upon visual information processing, in which the skill and knowledge of the map user are vitally important. The interpretation of the map also depends on understanding the meaning of the signs on the map, how these relate to the phenomena they represent, and the graphic structure in which they appear.

All human communication, in its widest sense, involves the use of signs. Language, mathematics, art, music – all employ signs either to express feeling or emotion, or to communicate information. The term 'communication' is frequently used to cover both, but it must be remembered that people can express themselves without necessarily intending to communicate, although the distinction between the two is often difficult to make. Generally speaking, we can accept that maps are mainly concerned with information. Whether or not a map can be conceived of as a personal expression, or whether it has any of the characteristics of a work of art, raises the question of aesthetics. It is a question which is common to all types of human creation which can be both functional and pleasing.

The problem of how signs, or systems of signs, operate, has been investigated by many philosophers and psychologists. Because of the great importance of language – the most important single means of human communication – and the close connection between language and thought, most analyses have concentrated on linguistic signs. Through language, people are able to refer to and think about both concrete objects and abstractions, past events and experiences, and affairs remote both in space and time. Indeed, how we are able to know, understand and describe the external world is a central problem in the branch of philosophy known as epistemology, which is much concerned with words and their meanings. Few analyses of sign systems make any reference to maps, and when they do they normally take a simplistic view and consign them to the same group as pictures and diagrams. So it is necessary to consider whether the signs used on maps have a systematic basis, and whether this is similar to, or specifically different from, other sign systems such as language.

All sign systems used to communicate information can operate so that they represent or refer to something which is not actually present; that is, they serve at the lowest level as a substitute for direct experience. As George (1962: 189) describes it, 'knowledge by acquaintance calls for direct experience with the relevant stimuli and responses, while knowledge by description involves indirect experience with these events through the mediation of signs, which, the organism has learned, refer to those same events.' Morris (1971: 189) makes a similar point when he observes that 'Through signs the individual directs his behaviour with reference to things and situations which he may never have encountered and never can encounter.' If we also take into account Piaget's comment (1971: 191) that 'one of the essential functions of knowing is to bring about foresight,' then it is clear that the map must also belong to these sign-using activities, because the map enables us to perceive real phenomena from a point of view beyond direct experience, and to anticipate future needs.

It is clear, therefore, that any item which is to be recorded or communicated must be affected by how the sign system works, and how efficient people are at using it. Agreement on a system of signs has to function through the medium of individual human brains. Consequently, a discussion of how signs on a map operate is essential to an understanding of maps and map use. Such a discussion needs to concentrate on those points which are of special importance for maps, and in particular to demonstrate both the advantages and limitations of maps as specialized forms of human communication.

The analysis of sign systems

Since the time of classical Greek civilization, philosophers have been concerned with how people think, and therefore with the sign system of language through which the mind operates. John Locke introduced the term 'semiotic' to describe the philosophical study of signs (the term is now more commonly expressed as semiotics). The relationship between signs and what they represent (their meaning) is the special field of 'semantics', and it is evident that this must be an important part of the analysis of any sign system. The term *'semiologie'* is extensively used in French, and although there seems to be no single or agreed definition, it is generally directed mainly towards the meaning of signs, and therefore overlaps semiotics.

Semiotics

The fundamental concern of semiotics is to discover how human beings are able to communicate with one another, and in this sense how meaning is conveyed through the use of language or any other sign system. Munro (1987: 115) clarifies this by contrasting the objectives of semiotics and aesthetics. He remarks that the essential difference between a semiotic and an aesthetic view is that 'the first takes the object as a sign of something else while the second takes it as pleasurable in itself.' Although this view might be disputed by many other writers, it focuses attention upon a central question: whether any sign system is simply an arbitary device in which the signs refer unambiguously to other things, or whether an arrangement of signs produces something which is different in nature from the individual signs themselves. In this respect, it emphasizes the importance of context and interpretation. Despite its initial basis in language, modern semiotics at least attempts to include all forms of communication and expression, and therefore has to take into account the arts, as well as exposition.

The rise of semiotics is associated with the works of a small number of philosophers; Saussure on the one hand, and the Americans Peirce and Morris on the other. Many others, such as Eco, Barthes and Langer, have both criticized and amplified these initiations. The work of Saussure is closely associated with the notion that language is the fundamental means of communication, and as such can be used as the 'model' for the analysis of other sign systems. Defining semiology (the term originated by Saussure) as 'A science that studies the life of signs within society,' Innis explained that in Saussure's view 'language, better than anything else, offers a basis for understanding the semiological problem; but language, must, to put it correctly, be studied in itself' (Innis, 1985: 34–35). Saussure distinguished beween the sign itself (the signifier) and what it stands for (the signified), and maintained that the connection between them in a linguistic sign is wholly arbitrary. Thus he separated sign in this sense from symbol, which he realized was never arbitary, for it indicates 'the rudiment of a natural bond between the signifier and signified' (Innis, 1985: 38). According to Culler (1976: 91), Saussure's view was that 'Semiology is based on the assumption that insofar as human actions or productions convey meaning, insofar as they function as signs, there must be an underlying system of conventions and distinctions which makes this meaning possible.' He also made clear that although linguistic signs depend on a linear succession, other types of signifiers, such as visual ones, can operate in several dimensions.

The American philosopher Peirce attempted a far wider examination of sign systems. His famous analysis of the three components of a sign system – the sign vehicle, its referential meaning, and its interpretation – was intended to provide a fundamental basis for all analyses of signs (Peirce, 1934). Although Peirce also accepted that the function of a sign is to stand for, or represent, something else, he explored at length the great variety of signs in an attempt to provide a full classification. His distinctions between indexes, icons and symbols have been used in many subsequent enquiries (see Hookway, 1985: 125). He regarded semiotics as essentially logical, in the sense that it could be applied systematically to provide a full and consistent account of all sign-using activity.

Peirce (1934) attempted to deal with all sign manifestations, including the use of graphic and pictorial signs, but like many other philosophers he appeared to regard maps as basically the same as diagrams. He described pictures and diagrams as essentially iconic, but appeared to be more concerned with the iconicity of the two-dimensional field of the diagram or picture than the specific signs of which it is composed. He also stated that 'Many diagrams resemble their objects not at all in looks; it is only in respect to the relations of their parts that their likeness consists' (Peirce, 1934: 159), which would seem to encompass the use of abstract signs in an iconic field.

His definition of a symbol as a conventional sign is interesting, bearing in mind that for a long period the legend of a map was often referred to as a list of conventional signs. He also regarded language as consisting of symbols, in marked contrast to Saussure. Perhaps his most useful contribution to the analysis of signs in relation to maps is the statement that 'A symbol, we have seen, cannot indicate any particular thing; it denotes a kind of thing' (Peirce, 1934: 169). In this the classificatory character of map signs is expressed, whether they take the form of icons or purely conventional devices. Iconic signs based on similarity were classified by Peirce as hypoicons, within which he distinguished images, diagrams and metaphors. Whereas images represent appearances directly, diagrams are concerned with relations, and those which represent one thing by a parallelism with something else are metaphors.

Although it is clear that Peirce regarded semiotics as essentially logical, he does on occasions refer to the aesthetic and emotive properties of signs. This relates the pragmatic dimensions of sign use to what Peirce termed speculative rhetoric, which he seemed to conceive as a examination of the way in which symbolic representation may be made effective (see Smith, 1972). Lyne (1982) explains that although Peirce's theory describes the types of sign effects that may occur, it does not take into account the fact that signs can only exist within a culture, not in isolation. He goes on to point out that even in scientific communication it may be necessary to present the truth in different ways to different people. This line of thought, of course, sees the use of signs as having a rhetorical dimension.

Peirce defined the six basic types of sign as signal, symptom, symbol, icon, index and name (see Greenlee, 1973). He developed this into a highly elaborate analysis, much of which is difficult to interpret. Nevertheless, this was a starting point for many further investigations, of which that of Morris (1971) is perhaps the most fully developed. It is proposed, therefore, to take the framework described by Morris, and see to what extent it can also serve as a basis for analysing signs used on maps, despite the fact that Morris makes no detailed reference to maps or other types of graphic image.

According to Morris, a sign system has three main components: the sign *vehicle*, the *designatum*, and the *interpretant*. The sign vehicle refers to the sign itself; in the case of a map it is the actual mark printed or displayed. The designatum is what the sign refers to, or what it is intended to mean. The interpretant is the means of the understanding,

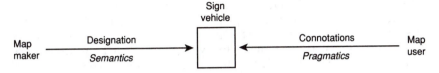

Figure 24. The elements of a sign system, and their relations

that is, the shared thought or concept. To take a simple map example, a blue line (the sign vehicle) may designate the feature category river, and the term 'river' is agreed in the English language to refer to a body of water flowing in a channel. In Morris's words (1971: 45), 'The interpreter of a sign is an organism: the interpretant is the habit of the organism to respond, because of the sign vehicle, to absent objects . . . as if they were present.'

Other writers refer to designation and connotation as the two main aspects of sign function, designation being what the sign represents, and connotation what the perceiver understands as the meaning (Fig. 24).

Morris goes on to distinguish between designatum and *denotatum*, designatum referring to the concept or category which the sign represents, and denotatum (denotata) to all the individual members of the category. The distinction is an important one in the analysis of map signs. Although the legend symbol may designate the concept 'river', the map includes particular and individual rivers belonging to that category. It does not follow that because a certain category is designated in the legend that all the denotata will appear on the map. In many cases, only a selection of the total denotata will be included, even though this is never expressly stated (Fig. 25).

The relationships between sign vehicle, designation and interpretation are described by Morris as syntactics, semantics and pragmatics. Syntactics refers to the ordered relationship beween signs.

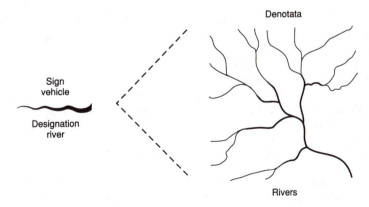

Figure 25. Designation and denotation

All sign systems must have rules which organize them. As syntactics is concerned with order and arrangement, then the map is clearly a special case, because any syntactical structure can only be presented through the locational structure of the map, which is itself a two-dimensional arrangement. This raises an interesting question. Does syntactical structure apply to the locational framework of the map, the signs representing the geographical graticule or grid, or does it refer to the geographical positions of the phenomena represented on the map? If the latter view is taken, then a map has no syntax in the ordinary sense. But if there is no syntax, why does a map need to have signs which only serve to indicate position? If syntax is regarded as applying to the locational structure of the map, then this would help to explain why it is so carefully defined through coordinate systems. On the other hand, if it applies to the geographical locations of individual features, then these exist quite independently of the sign system.

Semantics deals with the relationships between sign vehicles and what they designate, and therefore what they denote. A semantic rule, as Morris (1971: 36) describes it, is a rule 'which determines under which conditions a sign is applicable to an object.' It is in this area that there is the greatest degree of controversy and the widest range of terminology employed by philosophers, especially as semantics is often employed to refer to all aspects of meaning, both designation and connotation. The relationship of the sign to its designatum raises the question of the different types of signs, and the means by which they operate. An analysis of the types of sign used on maps is particularly important.

By pragmatics (not to be confused with the philosophical school of Pragmatism), Morris means the relationship between the sign and the interpreter, and in particular the problem of how the intended designation is understood by the perceiver. A sign system only functions because the creator and the user attempt to agree on the meanings of the signs and their relationships. Whether this can be fully accomplished in any sign system is debatable, but a great deal of education is directed towards the goal of ensuring that signs are correctly interpreted and understood. On this point Peirce concluded (see Greenlee, 1973: 99) that 'nothing is a sign unless it is interpreted as a sign.' In terms of a map, the perception of a blue line does not lead to the visual response 'a blue line' but to the identification 'river'. As Rescher (1973: 100) puts it, 'The very concept of identification involves some reference to the directable attention of a comprehending intelligence.' Whether or not the map maker and the map user have exactly the same concept of 'river' is a neglected but significant point.

Both semantics and pragmatics are deeply concerned with concepts. Although the designation of a map sign may be determined by the map maker, the interpretation of that sign is a function of the map user. Because of the apparent simplicity of many map signs (river, boundary, contour), the importance of interpretation is often overlooked. Yet it is significant enough to provide a basis for a classification of map types.

Signs and symbols

Although Peirce classified signs into six main types, most writers emphasize the distinction between two main groups, signals and symbols. A signal requires a single, predetermined response. It is not open to various interpretations, and does not represent the characteristics of an object. In the other set, Morris places symbols, icons and indexical signs. Symbols are regarded as characterizing signs, and they can be subdivided according to the nature of this characterization, i.e. what aspect of the designatum they represent. An icon represents the appearance of something, and is thus directly representational. On the other hand, an indexical sign (indicator) 'designates what it directs attention to' (Morris, 1971: 37). It does not represent an object as such, but indicates a certain place or position. Although this is commonly applied to the grammatical signs of verbal syntax, it also fits those signs used on maps which serve to indicate position (the geographical graticule and the grid), which do not refer to any actual phenomena, but organize and arrange the space in which they appear.

Ducasse (1968: 80) makes the same basic point when he distinguishes between indicative and quiddative signs. The function of an indicative sign is 'to orient our attention to some place in an order system,' whereas that of a quiddative (or descriptive) sign 'is to make us conceive a certain kind of thing.' The same entity may function as both.

Greenlee (1973: 135) argues that although it is possible to distinguish between icon, index and symbol in the sense of what they represent, it is difficult to draw exact boundaries between them according to the way they function. In this respect he tends to regard all of them as symbols – icons and indexical signs being specialized subclasses.

Symbols and representation

The manner in which a symbol represents or characterizes its designatum is a subject widely discussed. Many writers, including Morris, argue that whereas an icon is a likeness (as in an image), a symbol has only a conventional relationship with its designatum. Whether or not a map is iconic also depends on what aspect of the map is being considered. Regarded as a whole, the two-dimensional extent of the map is iconic because it represents directly the two-dimensional 'surface' of the real world, although strictly speaking this makes it equivalent to a plan, not a map. In one of the few direct references to maps, Peirce's view on this is expressed by Greenlee (1973: 70): 'Maps and diagrams are typical of the illustrations Peirce gives for icons.' In going on to distinguish a further property of the icon as a sign (which applies particularly to a map or plan), Peirce refers to the fact that 'by the direct observation of it other truths concerning its object can be discovered than those which suffice to determine its construction' (Greenlee, 1973: 78). This of course refers to the map's property of demonstrating spatial relations, by which the whole is greater than the sum of its parts. Greenlee (1973: 9) refers to this as the icon's exhibitive import, which 'consists in the power of the icon to help direct attention to other properties of the object than those the sign-vehicle is conventionally interpreted to represent.' It is also the point made by Bertin (1978: 121) when he defines a graphic as 'an image that transcribes relations between elements or groups of elements.'

Because the symbol depends upon a convention, i.e. an agreement between the creator and the user, there is no limit to what it can represent, and it can deal perfectly well with abstractions and emotions, as well as 'real' phenomena. According to Sebeok (1976: 134), a symbol is 'a sign without either similarity or contiguity – but only a conventional link between its signifier and its denotata.' The same point is made by Crick (1976: 135): 'The term symbol, of course, implies a relation of "standing for" rather than one of reproduction.'

This viewpoint can be seen in the old cartographic tradition of referring to symbols as 'conventional signs'. Yet the use of symbols on maps is more varied and complex than this would suggest. For example, the representation of a buoy on a chart, with its shape, topmark, etc., is certainly meant to be iconic, to represent its physical appearance in a stylized way. A miniature runway diagram representing an airfield has something of the same character on an aeronautical chart. Yet a black line representing a fence on a large-scale plan cannot be said to represent the appearance of a fence in any way, and indeed it might be impossible to deduce the description 'fence' at all. On the other hand, a black square representing an isolated building on a topographic map is not wholly 'conventional': its regular, angular form has some element of a typical building plan.

Although the wording is involved, Akhmanova (1977: 9) takes up this problem when she states that 'We speak of a symbol when the choice of signans is not wholly arbitrary or conventional, without, however, being similar to, or resembling the denotatum in the ordinary sense of the word.' The qualification is important for map signs, because they can and do represent attributes of objects apart from physical appearance. As Miller (1973: 104) says, 'It is precisely the advantage of maps that they can use many combinations of pictorial and abstract devices to serve a wide variety of purposes.' It would seem preferable, therefore, to take a broad view of map symbols, extending from iconic at one extreme to purely conventional at the other (Fig. 26).

Symbols and concepts

In order to refer to the host of individual features of the external world in a comprehensible manner, they have to be ordered and classified into groups, the members of which have certain attributes in common. This process of classification is essential to language and thought, and operates through the use of concepts. Terms such as river, boundary, built-up area and road all indicate a large variety of individually different features which can be classified or categorized into groups on the basis of shared characteristics. These classificatory concepts are

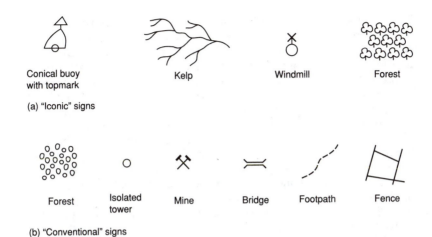

Figure 26. Types of sign: (a) iconic, and (b) conventional

essential to the process of description, but the term 'concept' is used in so many ways that it requires some examination.

Johnston (1972) distinguishes between singular concepts, conceptions, and classificatory concepts. Other writers refer to general concepts and individual concepts, and the difference is often expressed through the use of the personal adjective. The concept 'continent' is a general classificatory concept, held in common by people who share the same language. My 'conception' of the continent of Africa is personal and individual. Some philosophers, such as Geach (1957: 13), use the term only in this sense of conception: 'A concept, as I am using the term, is subjective – it is a mental capacity belonging to a particular person.' Although the use of the two terms is confused, the distinction between the two meanings is important. The information shown on a map must depend on the use of shared concepts to make the symbols intelligible, and not on personal conceptions. The point is important also in considerations of learning and mental development. Klausmeier *et al.* (1974: 4) make the distinction between mental constructs of individuals (conceptions) and 'identifiable public entities that comprise part of the substance of the various disciplines.'

Concepts are developed through the use of language, and education is largely directed towards their attainment. The ability to conceive and use abstract concepts is affected by age and education, and there is an important stage in a child's mental development when abstraction becomes possible.

As concepts are learned, they can be vague, incomplete or inadequate. Consequently they can also be sources of error. As Miller (1973: 125) puts it, the use of concepts 'does generate certain mistakes because of the superficial resemblances between things really different and because of the placing in the same class of things that turn out by further experience to be not equivalent.'

The designation of a symbol, therefore, which is a semantic relationship between the sign and its object, depends upon the use of a concept which is, or is assumed to be, shared by both the creator and the user. In a map legend, the symbol 'canal' refers to all those features which are recognized as belonging to that class or category, and it is assumed that both the map maker and the map user will share the concept 'canal'. If this is not so, then communication is either severely restricted or impossible.

The attributes of a feature or phenomenon which determine the class to which it belongs are defined as the concept's intension, and all the individual features or occurrences which fit into the class are called the concept's extension. The parallel with Morris's terminology of sign function is clear. The concept's intension is equivalent to what the sign designates, and the concept's extension is what the sign denotes. Generally speaking, if an intension is increased (made more particular), then extension is decreased.

According to Nørreklit (1973: 92): 'By a concept's intension we understand the distinguishing features or criteria an object must satisfy in order to belong to its extension. On the other hand by an object's extension we understand the class of all objects that satisfy these criteria.' The relationship between concept and sign is also expressed by Morris (1971: 37): 'A "concept" may be recognised as a semantical rule determining the use of characterising signs.'

Understanding the use of general or classifactory concepts in this sense is fundamental to understanding maps. It must be appreciated that the signs and symbols on a map do not describe individual features – they are represented through general concepts as belonging to classes or categories. The unique information given about any feature (even if the category is subdivided in detail) is its location, and consequently it is the provision of locational information that is the primary function of the map. The point is made by Petchenik (1977: 121) in saying that 'Cartographers are not concerned fundamentally with the nature of objects *per se*, but rather with a particular set of relations among those objects.' In this respect, the concepts through which features are described are a means, not an end.

The use of general concepts also underlines the fact that a map is not a 'picture' or an 'image' of reality. Direct perceptions of the external world can only be of particular things, and a visual image can only be of one particular object. We do not see the category 'trees' in the landscape: only individual and particular trees. As Nørreklit says (1973: 88), 'It does not lie in the nature of images to denote many different objects.' The use of concepts depends very much on language, and it is largely for this reason that most philosophers regard language as fundamental to all types of human communication. It is also evident that the concepts through which the map information is expressed must be known and understood before the map can be used. Although a symbol may be correctly identified by reference to the map legend, or 'conventional signs', the map cannot be interpreted until the map user comprehends the meaning of the symbol, that is, the semantic relationship between the sign and what it designates.

This comprehension is affected by experience, learning and expectation. In the three map examples given in Chapter 1 (see Figs 1–3), no legend is included, and it is assumed that the maps can be understood on the basis of the general description of what the maps are about. This therefore relies heavily on the reader's acquaintance with what sort of map symbols are likely to occur, and the relationships between them. Nevertheless, it is equally probable that these simplified maps would be unintelligible to a person entirely unfamiliar with maps of any sort. This demonstrates how common it is to make assumptions about the knowledge and experience of readers and their familiarity with concepts.

Modes of signifying

After dealing wih types of sign, Morris goes on to analyse three other aspects of sign function: sign mode, sign usage and sign adequacy. Although much of this is concerned essentially with linguistic problems, it also raises points which further clarify the use of signs on maps.

In modes of signifying, Morris is concerned with the ways in which signs operate. He makes a basic distinction, and one introduced by many philosophers, between signs which inform or state, and those which are emotive. These two groups are also referred to as cognitive and non-cognitive, or referential and expressive. For example, the statement that a certain feature is in a certain place is normally intended to be informative, and is not expected to arouse any emotional reaction. The statement that one thing is better than another is a valuation or judgement, and depends on the view of whoever makes it. Informative statements can be verified; judgements or appraisals cannot.

Morris distinguishes between signs which are identifiors, designators, appraisors and prescriptors. Identifiors are concerned with location, designators with discrimination, appraisors with valuation, and prescriptors with obligation. It is immediately clear that all map symbols are identifiors, in the sense that they all point to a place in the locational structure. In Morris' words (1971: 154), 'An identifior then, signifies the location of something.' It seems that they must also be designators, as they enable the map user to discriminate between one thing and another, although in some cases only to a very limited degree. According to Morris (1971: 156), 'Designators, as signifying discriminata, signify to an organism the characteristics of what it will or might encounter; they do not signify the import of that object for its goals.' Therefore, designative identifiors correspond in general to what we may call statements, and they are specially adapted for the primary purpose of giving information. Salomon (1966) makes a similar analysis in distinguishing between informative, prescriptive and emotive signs.

Sign usage

In dealing with sign usage, Morris contrasts those signs which are employed to inform about, to show different values in, to incite or arouse, and to influence behaviour about other signs. This repeats the distinction made in sign mode between informative and valuative or emotive signs. There is no reason to doubt that signs on maps are basically informative, but it is possible to find examples where they are valuative, emotive or prescriptive. A tourist map may include a symbol for 'scenic route', which must be a subjective valuation. A chart may have a strongly marked line identifying a 'prohibited zone'. Although the sign itself is an ascriptive designator, it is also an injunction about behaviour at a certain location, rather than just a description of a thing or event. Although the appraisive or prescriptive function of a sign may not be intended, it can cause confusion in map use. The obvious example is in the representation of international boundaries. Although the map symbol is only intended to make a statement, i.e. that a certain boundary occurs in a certain place, it can be interpreted as appraisive from a political point of view. Maps can also be devised to be emotive rather than informative, relying upon the fact that the habit of the map user to regard maps as informative can be exploited to achieve an emotive effect. The maps produced for war propaganda are the classic examples, but this exploitation of appraisive sign use can be found in many maps, especially where they are intended to encourage a certain behaviour, such as tourism.

Sign adequacy

In his treatment of sign adequacy, Morris defines adequately informative as convincing and adequately valuative as effective. This in turn raises the question of the reliability of the information, its accuracy, and the acceptance by people that informative statements can be held to be true or correct. In Morris's words (1971: 178), 'A sign is informatively adequate (or convincing) when its production causes an interpreter to act as if something had certain characteristics.' So far as concepts are concerned, this is relatively straightforward, but a map user must be able to rely upon the correctness of the locational information as well.

If the information in a map is to be relied on (i.e. regarded as convincing), and if the map consists mainly of signs which make statements, then these statements must be regarded as matters of fact. According to Russell (1914: 51), 'The existing world consists of many things with many qualities and relations. "Fact", has two or more constituents. When it assigns a quality to a thing, it has only two constituents, the thing and the quality. When it consists of a relation between two things, it has three constituents, the things and the relation.'

The implications for the informational structure of a map are clear. Not only must the phenomena represented be correctly classified, but their relations in space must also be assumed to be 'correct' by the map user. That a certain city is in a certain place on the Earth's surface – a statement made in map example A (see Fig. 1) – is a statement encompassing two things and a relation. As such it must be interpreted as a statement of fact, and in Russell's words (1914: 52): 'The fact itself is objective and independent of our thought or opinion about it.' The same point is made by Sprigge (1970: 84): 'There is not something, which if it is true is a fact, and if it is false something else on the same level.'

It can be argued, of course, that maps can show 'theories' or hypotheses about phenomena. But the fundamental structure remains the same. The positions indicated by the map – which may only be what a certain author proposes – represent what the position would be if the statements were true.

If a map can be used in direct comparison with real phenomena (as a large-scale topographic map can be used for orientation in the field), then the evidence of the map can be empirically tested as to what is verifiable. But very many maps contain information which cannot be directly verified by the user, and therefore what they state must be assumed to be true. Because the specific information in a map deals with location, then the reliability of the locational information is crucial. Generally speaking, the user must rely upon it as indirect evidence, which can only occasionally be tested against the direct evidence of personal experience. The importance of any doubts about the reliability of map information can be illustrated by the reactions of map users who discover that something on a map, to their knowledge (or belief), is in error. Despite the fact that the particular erroneous item may be surrounded by hundreds of statements that are true, doubts about the 'accuracy' and therefore the reliability of the map persist.

The passionate attachment of many of those who provide information for maps, especially surveyors, to 'accuracy' and 'precision' is often slighted by map users (and indeed some map makers), who regard their preoccupation as extravagant or over-indulged. Yet it is clear that the surveyors are motivated, even if subconsciously, by an instinctive understanding that if the locational information is open to question, the map cannot be considered as 'factual', and the whole function of the map is thereby reduced. From a philosophical point of view it may be impossible to make any absolute distinction between what is true and what is false. Nevertheless, the use of signs on a map is based on the assumption that the phenomena represented are located where they are stated to be.

It is curious that although maps can only function through the use of signs, so little attention has been given to their characterstics and organization. Yet a basic understanding of how signs on a map operate is essential to both making and using maps, and a proper appreciation of this would remove a great deal of confusion both in describing maps and in discussing problems of map use. In particular, the distinction between what the map actually states, and what the user interprets, is often misunderstood.

Indexical signs and representation

The problem of how to classify those signs on maps which do not refer to or characterize concrete objects but only places or positions has already been raised. A meridian of longitude has no existence, and therefore no character or appearance, so it is essentially different to a river or a road. Many boundary lines, especially those in internal administration, also have no physical presence, even though they can be located. These, however, can be characterized in a different way, because they do affect human activity, movement or organization, and in this sense are 'prescriptive' signs. But a line which is part of a coordinate system has no direct effect or consequence; it simply attaches a value to a position. Any coordinate system is simply a means of referring to points in space, and therefore such coordinate systems can be regarded as sets of indexical signs.

The relationship between sign and designation becomes more complex when the representation of real objects or surfaces is done indirectly. A contour is not a feature, and has no individual existence. In this respect it is a measured line with a value, similar to a meridian or parallel. But by means of such lines, or combinations of points and lines, variations in shape of real phenomena can be interpreted, even though contours themselves do not designate or characterize real features in the terrain such as hills or valleys. In this respect the contour is an identifior, but at the opposite end of the sign spectrum from the iconic symbol. Because the contour (and the spot height) can 'place' things in the third dimension, it is similar in some respects to the indexical signs which 'place' things in a two-dimensional field. But in this case the positions of the lines are dependent upon a 'real' surface, and not just an arbitrary reference system.

The other sets of signs that are difficult to categorize are the signs used on many plans, which do not have specific designation. On most street plans, for example, there is no explanation in a legend of what the lines 'mean' (Fig. 27). In this case they also point to a position, but only 'indicate' that something changes along the position shown by the line, i.e. that what is on one side is different to what is on the other. What is represented may be identified either by direct observation on the ground, or imagined on the basis of previous experience and expectation. In both cases the burden of correct interpretation is placed on the map user, for a sign that simply points to 'something' is operating at the lowest level of representation.

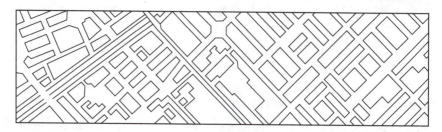

Figure 27. Identification and the typical street plan. What do the lines 'mean'?

Place names as map signs

In the introductory analysis of signs, it was pointed out that Peirce regarded names as one of the six main sign types. On the other hand, Morris treats them as a subset of identifiors, along with other indicators. Geographical place names are not easy to classify on this basis. They are not purely indexical signs, even though they point to a location. Some of them have elements of characterizing signs, because a name such as 'Red River' may be descriptive. They are certainly conventional because they can be changed at will, and depend solely on agreement in usage.

Geographical names must be treated as a special case. Although they are abstract, and do not necessarily signify any attribute of the object or feature, they serve to refer to particular places, not just points in a coordinate system. Therefore, they act as a connecting link between the two sign systems of the map – indexical signs and characterizing signs – because unlike the coordinate reference systems they refer to specific features in the map content. The spatial reference of the name 'Africa' is quite clear: it is very difficult to replace by a set of coordinates. Coordinate systems are systematic and precise. They faciliate exact determination of position, distance and direction, and are equally accessible to all map users who are prepared to learn about them. Place names are unsystematic, arbitrary and frequently ambiguous. They are unevenly distributed, repetitious and incomplete. They are only valid amongst users who are familiar with them. They are circumscribed by the organization of a particular language and script, and although they are 'linguistic' cannot function through translation. They pose many inter-linguistic problems. They depend very much upon familiarity in a community, and they are the most convenient way of referring to entities that have a geographical location. It is probably true that a very common use for maps (as shown in map example A; see Fig. 1) is to establish the location of a place first learned about by name. The connection between a name and its reference feature has to be learned, and for places outside direct experience the map is the most important source of indirect evidence.

Not surprisingly, names present endless problems to map makers. Even within small areas they are subject to change, new ones being added and old ones disappearing. The definition of the 'correct' name for a place or feature is often a subject of controversy. Yet it is clear that the names of places in the immediate environment are an important element in a person's consciousness of self-location, and often because of this, changes to place names in use are often strongly resisted. For most people, to 'know where you are' is almost certainly expressed by reference to named locations.

Names on a map, therefore, are a system of signs, which have some of the characteristics of both indexical and descriptive signs. They can be emotive, appraisive, and even prescriptive (e.g. Gallows Hill, Danger Bay, or Crazy Lake). They can only operate in conjunction with the other sign systems which point to locations and identify features.

Despite their arbitrary nature, they are an essential means of place identification, and this is reflected by the importance given to them on maps.

Sign designation and map type

The characterizing signs or symbols used on a map designate specific features or phenomena through the use of concepts. They can also provide a useful basis for distinguishing different types of map. In what can be called 'general' maps, or 'multi-purpose' maps, regardless of scale, it is assumed that the concepts designated by the map symbols will be understood by the map user without any special knowledge, on the basis that the map user is an adult of average intelligence and normal education in a particular society. Ordinary geographical, social and political terms are used to describe the map content. Whether or not this is always adequate is debatable, but certainly the assumption is made that 'built-up area', 'motorway' and 'contour' will require no further explanation. On the other hand, various types of specialized map require an understanding of concepts which are the products of particular disciplines, either for their connotation or their interpretation. The map may be concerned with a special subject, such as geology, or it may be for a special purpose, such as navigation. In both cases, the degree of specialization will affect both the locational structure employed and the concepts designated by the symbols (Fig. 28).

GENERAL	SPECIAL PURPOSE	SPECIAL SUBJECT
Major road	Non-directional radio-beacon	Felsite
Canal	Group occulting	Porphyrite
Bridge	Chart datum	Camptonite
Aqueduct	Drying height	Diorite

Figure 28. Sign designation, concept and map type

There can be different levels of specialization. A small-scale map of the geological formations of a country may contain some designations which include terms familiar in everyday language, such as sandstone or limestone. But in addition, it is likely that classifications such as 'Silurian' or 'Precambrian' will also occur, which are unlikely to be fully comprehensible to the map user unless he has some specialized knowledge of geological concepts. Specialization of subject matter in this sense covers a wide range. What is significant is that the concepts must be learned before the map can be fully understood. These concepts are not the product of anything specifically 'cartographic', and they are not part of any cartographic theory.

The second specialized group consists of special-purpose maps.

Although they are likely to contain some specialized subject matter, the selection of this is due to the need to satisfy certain map-using activities. In practice, the vast majority of such maps are devoted to navigation and orientation. They range all the way from the familiar road map, which may have much the same content as a topographic 'general' map of the same scale (but designed with a different set of priorities), to an aircraft instrumental flight chart, devised to assist quite specific navigational procedures. The other main subgroups consist of those maps devoted to planning and forecasting, their purpose being to facilitate decisions within specialized fields.

The specialized map may also require a specialized structure in order to provide for a particular map use. The functional implications of a 'Mercator' chart are necessarily apparent to a navigator, but not normally evident to a layman. In the case of a navigational chart, on which bearings and courses are plotted, the navigator has to be fully aware of the consequences of scale and projection. For less specialized maps, such as road maps, where the user's requirements are less critical, difficulties are more likely to arise from failure to appreciate the factors of scale and generalization. In broad terms, maps designed for specific uses are more likely to demand a specific locational structure, and therefore a higher level of cartographic understanding by the user.

As Robinson and Petchenik (1976: 109) point out, there is no simple or clear line of division between various types of maps. General topographic maps shade into slightly specialized 'tourist' maps, or 'road' maps, and the object is to satisfy special interests rather than highly specialized requirements. But, in basic terms, the understanding by the map user of a map's form and content is a function of his comprehension of the map's sign structure, in both its locational and descriptive aspects.

CHAPTER 5

Representation in two dimensions

Three dimensions into two

Maps, pictures and diagrams are often grouped together, as all three have the fundamental characteristic of being two-dimensional. Although Goodman (1969), like Peirce (1934), appears to regard maps primarily as diagrams, maps, of course, can include both pictorial and diagrammatic elements, even though the fundamental structure of the map is distinct. In dealing with maps, this introduces a difficulty which parallels that generated by the concentration on language as the chief means of expression and communication. Most conceptions of the two-dimensional graphic image are dominated by the concept of picture, and so the essential distinctions between map and picture are often blurred. Even accounts of cartographic principles make use of the analogy of the map as a 'picture' of the ground. What maps and pictures have in common is that both are analogue devices, and both depend on the acceptance of conventions. In both the representation is continuous, just as three-dimensional space or the Earth's surface is continuous. This is the fundamental contrast with linguistic sign systems which have a linear arrangement. The point is made by Goodman (1969) and amplified by Pitkänen (1980), but with maps the notion that the representation is continuous needs to be treated with some care. In pictorial linear perspective, where position in the picture space corresponds to a perceived view, real or imaginary, there are no 'empty' spaces or voids. With a map, the plane may represent no more than notional surface, on which particular objects are located.

In viewing pictures, a set of conventions, or agreed assumptions about the appearance of objects and the treatment of pictorial space, is essential, even though they are not overtly described. Most members of any culture learn them through acquaintance, like speech, and they lie at a subconscious level. People generally learn to use maps by being involved in a map-using activity rather than by formal study, and normally without questioning the conventions on which maps are based. It is only when such conventions are suddenly interrupted that they attract attention. When the Cubists combined several different viewpoints in a single picture, people were astonished, even though the use of more than one viewpoint, and therefore perspective, had been employed in both pictures and maps for more than three millenia.

Part of the problem of making comparisons between maps and pictures is the need to disentangle two distinct aspects: the representation as a whole, and the representation of particulars. For both maps and pictures, the two-dimensional bounded surface is iconic, that is, it can serve to represent directly certain aspects of a three-dimensional world within the limitations of two dimensions. In this respect the map is iconic, because one surface can be represented by another, despite the distortion involved. Unlike a picture, the particular elements in the map can be, and frequently are, interpreted individually. The major part of the content is likely to consist of discrete signs. A picture, like any work of art, only makes sense as a whole; a map can be used as a complete representation, or in part, or for particular items. Even the notion of 'three-dimensional' has different connotations. In addition, maps cover a far greater range of scale and area than pictures, and therefore make much more complex demands on projection than normal or conventional pictorial representation.

Although an oblique view of a continent or a hemisphere may mimic the perceptual characteristics of normal vision, and therefore of a picture, it still has to make use of the conventions of the map. A map can be based on concepts and uses of projection that are not limited by what can be directly perceived, which of course is one of its great advantages.

For maps, three-dimensional has three connotations. As the Earth is nearly spherical, this continuous curved surface has to be represented on a two-dimensional plane – the classic problem of map projection. From the point of view of the user, there is a considerable difference between large-scale individual maps of small areas, and small-scale maps of large areas. For the former, the correspondence between a small part of the spherical surface and a map plane is close: for the latter it introduces enormous distortions, which can have only conventional solutions. Strictly speaking, it is impossible to geometrically project more than one hemisphere directly onto a plane, and therefore representations of the whole sphere as a single continuous image require conventions of a different nature.

The second aspect is that the surface of the Earth is three-dimensional in that it varies in relief, and it is this irregular surface which has to be represented by the map. Of course it can be ignored, in which case the representation is a plan; that is, it deals with all particular objects as though they actually did occur on a plane. The variation in elevation of the relief can be described, and it can even be suggested visually by tonal modelling equivalent to that used in drawing or painting, but this must apply to the whole of the map surface, not only that apparent from a single viewpoint. Pitkänen (1980: 42) points out that 'a map, typically, reduces three-dimensionality to two-dimensionality in a way a picture, typically, does not.'

The third element is that three-dimensional objects can be represented on a map as part of the continuous surface, or as features occurring individually on it. Where the three-dimensional nature of the object is regarded as significant, as is often the case with buildings or tall vegetation, then these may be presented iconically, but the extent to which this can be done is usually limited, because of the space required by iconic representation.

In this respect it is customary to discuss 'iconic' primarily in relation to the classification of specific map symbols, in which the superficial appearance of the object may form the basis of the symbolic representation. There are several different ways in which individual three-dimensional objects can be projected. Indeed this is one of the links between map and picture. The map as a whole is iconic in a different way, because projection is used in a manner which is distinct from that of the picture or the diagram.

Projection and correspondence

For human existence and activity, knowing 'where things are' is essential, and over a limited distance being able to place things in a two-dimensional sense is adequate for most purposes. Egocentric concepts of distance and direction in two dimensions can be amplified mentally by taking into account the approximate relief of the land where this is important. Characteristically, the inhabitants of mountainous regions think of distance in units of time rather than in units of length, but most people 'place' things in terms of an imaginary horizontal plane. It also probably accounts for the fact that the flat map of the Earth looks perfectly sensible, even though it is based on an enormous distortion. The difficulty of appreciating the distortion present on maps of the whole Earth, or large parts of it, is responsible for much of the misinterpretation of geographical relationships that is all too possible. The correspondence between a two-dimensional representation of the Earth and the Earth itself is extremely limited. It can only operate at all through a conventional arrangement. The representation is an artificial construction on which an interpretation can be performed, but only by understanding or accepting the conventions which underlie it.

Map projection

As maps are concerned fundamentally with showing the location of things in relation to the Earth's surface, they are conceptually based on what Hagen (1986) terms 'metric' projection. In this, the representation of the third dimension is sacrificed in order to maintain correct angles and distances for two dimensions. The essential difference between map and picture is that whereas the picture is concerned with three-dimensional pictorial space as a representation of a field of view, the map has to represent the three-dimensional surface of the spherical Earth on a plane, and this curved surface cannot be represented without

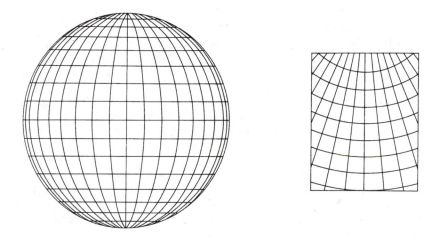

Figure 29. Transformation from the spherical Earth's surface to the map plane

distortion. A landscape painting of a particular scene is only 'correct' to the extent that it conforms with the rules of a given perspective; and a representation of all or part of the Earth's surface in a map is only 'correct' within the limits imposed by a particular projection.

The transformation from the curved surface of the Earth to the plane of the map (Fig. 29) is the function of projection, and at higher levels, of geodesy. Such transformations are calculated mathematically, and therefore the degree and distribution of distortion in a given projection, for a particular area and at a specific scale, is known. Generally speaking, projection problems have to be solved by the map maker, and in many cases there is no real need for the map user to be aware of the difficulties. The exception, of course, is if the map is to be used for a particular purpose, such as navigation, in which case the user must understand the effects of the projection on distance and direction. The limitations of a particular projection are understood by the map maker, and can be expressed at least approximately through the projection description. For example, the projection used in map example A (see Fig. 1) is a zenithal equal area. The facts are clear to the map maker, but not necessarily to the map user. For the general reference purposes that must be assumed for this particular map and scale, this may not be of any importance. But if the user attempts some action not consistent with the scale and projection – such as trying to measure the distance between two places – then the limitation becomes important, because it would involve a use not commensurate with the projection. A thorough account of the problems involved is given by Maling (1973).

The very much larger scale, and correspondingly smaller area, of map example B (see Fig. 2) reduces the matter of projection to a much less significant factor from the user's point of view. As this is only a small section of a map composed of many sheets covering a whole country, the map maker has a projection problem that is not apparent to the user. But within such an area, the relative positions of places will be so close to 'reality' that the map user can regard them as correct. Any difficulties in measurement will arise from the scale, the limited precision of visual estimation, and the degree of generalization.

For a map or plan of a very small area it is possible to treat the section of the Earth's surface locally as a plane. A two-dimensional Cartesian grid is sufficient to enable the correct representation of angles and distances. The measurements made by the topographic surveyor on the actual surface have to be reduced to those which would correctly indicate dimensions and positions if the objects and features actually did lie on a plane.

For maps of larger areas this simple solution is inadequate, and the inevitable distortion between sphere and plane has to be dealt with systematically in some way. Although there are many geometrical projections of a spherical surface onto a plane, in practice the projections used for topographic maps are carefully devised so that angles at any point are correct, and the scale is the same in both directions. In order to minimize distortion, and so to control it within reasonable limits, the actual scale error is distributed over the map so that it never reaches more than a certain level. Maps of larger areas are produced by repeating the projection in bands or zones, so treating limited areas consistently, and keeping the maximum error to within a controlled amount. Such maps normally comprise many individual sheets in order to cover a large area at a comparatively large scale. As such they are the largest images produced in graphic form.

Smaller-scale maps of large areas, such as continents, encounter a much more difficult problem. Although they are produced with a nominal scale ratio, such as 1 : 5 million, this scale relationship is only correct from certain points or along certain lines. As it is no longer possible to restrict distortion in any one map sheet to within a level that does not affect normal map use, a choice has to be made as to which property of correctness should be preserved. So a map projection can be chosen to maintain correct angles and therefore shapes, or correct areas, or correct distances from one point, or use some intermediate compromise. What it cannot do is maintain correct dimensions throughout the entire map. As with pictorial representation, gaining one correspondence means sacrificing another.

Any form of geometrical projection of a spherical surface onto a plane is necessarily limited to one hemisphere or less. Therefore maps which include more than one hemisphere, or the whole of the Earth, cannot operate on the basis of normal projection. The projections used for such maps are synthetic constructions, based on accepted conventions. Because both topographic map series and small-scale topographic maps of large areas are likely to indicate the geographical graticule in the form of meridians and parallels, which themselves are the basis of the projection, the different nature of projection which becomes important in maps of more than a hemisphere is not visually evident to the map user. The same graphic device moves imperceptibly from a 'correct' representation to a highly artificial convention.

Representation of the sphere

One hemisphere or less can be projected directly onto a tangent plane, or indirectly through a tangent or intersecting cone or cylinder. The projection may be developed from a point lying in the centre of the sphere (gnomonic); from a point on the opposite surface of the sphere (stereographic); or from a point infinitely distant (the orthographic), that is in parallel. Although these relatively simple geometrical projections have their properties, and therefore distinct uses, in most cases projections of large areas are constructed in order to minimize the distortion of shape or area, or find a suitable compromise between them.

Once the map is used to represent more than a hemisphere the problem of representation becomes more difficult, for there is no means by which a spherical surface can be shown as a whole on a single plane without interrupting it. Although in many pictorial representations it is sufficient to show one side of a symmetrical object (such as the human figure), this is insufficient in a representation of the whole of the Earth's surface. The Earth, unlike a geometrical sphere, does have an axis of rotation, making possible a sense of cardinal direction, and an Equator lying midway between the poles. Apart from this there is no direct starting point in the choice of how the 'projection' is to be placed in relation to the sphere, apart from the arbitrary one of longitude.

It has often been stated that the only satisfactory representation of the Earth is a globe, retaining its three-dimensional character. It is also clear that in using a globe only one hemisphere is visible at any time, however often the globe is rotated. The desire to construct a representation which presents the whole surface in one view is therefore a perfectly natural one, despite its complexities. A world projection cannot be considered as a case similar to any of the projection methods described so far. It can only be a conventional device, in which continuity is sacrificed for completeness.

Any representation of the sphere in a single two-dimensional plane must combine information about both hemispheres. In so doing it both duplicates and divides. In combining two viewpoints it operates on the assumption that the map user understands what has taken place, and will react accordingly. The pictorial assumption that the viewer regards the representation as a continuous whole makes this difficult. Whatever method of projection is used for a world map, and however much ingenuity is exercised in maintaining correct shapes, or correct areas, one meridian is repeated at the two sides of the representation. The degree of distortion can be enormous (as, for example, where each pole is shown as a line as long as the Equator) but this can be comprehended by the user. What is more insidious, because it is less obvious, is that areas which are in fact in proximity on the Earth's surface are shown as being separated by a great distance.

In the projection most commonly used in maps made in Europe, where Europe lies more or less in the centre of the projection, the extremities of East and West are apparently far distant. The concepts so

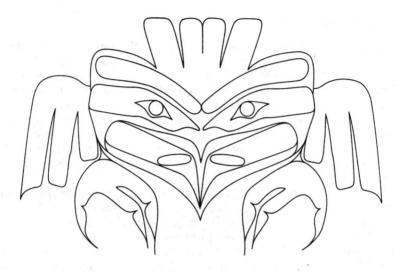

Figure 30. Split-field figure. Outline of a representation of a thunderbird in West Coast American Indian art

engendered are deep rooted. The modern notion of 'West' in distinction to 'East' is essentially derived from the fact that the western hemisphere on the world map is to the left of Europe, and the eastern hemisphere to the right, suggesting two extremes. Yet 'western' North America is closer to the East than it is to Europe. If the principle of defining zero longitude as the Greenwich meridian is followed, then of course most of 'western' Europe itself lies in the eastern hemisphere, not the western. The terms themselves are a fine example of the difficulty of assigning a specific meaning to a word, for they are loaded with historical, cultural and political implications. Similarly, if the Americas are placed centrally in the world map, then Europe and Africa become separated from much of Asia.

It is interesting to note that the idea of representing both sides of a three-dimensional object has occurred in other cultures and for quite different purposes. The so-called split-field composition of animal figures in North American West Coast Indian art has long been known, and it provides an interesting comparison with the standard map of the world. The animal figure, of course, may have the same side profiles but differs from front and rear. The highly stylized representation is constructed, therefore, so that the two sides are presented frontally, joined along a centre line (see Fig. 30). In a general sense, the two outer outlines are equivalent to the meridians which form the eastern and western extremities of a world map. Like a map of the world, they are synthetic representations of what is known, rather than what can be perceived in a single view.

Not surprisingly, comparisons between different projections for maps of the world, and the effects – intentional or otherwise – they have on map users, have a long history of argument and disagreement. On the one hand, the relative merits of this projection or that are presented; on the other, apparently unlimited 'new' projections are proposed. In some ways it is like a mathematical puzzle, in which there are endless permutations but no absolute solutions. Given the marked differences that arise from different projections, it is not surprising that claims are made that some world map projections reflect deliberate political and cultural bias. The whole problem provides an excellent example of the fundamentally rhetorical nature of maps, for with maps of the world it is impossible to separate the information from the representation. It is certainly possible to manipulate the projection to influence the way in which the world will be perceived.

Any map of the world can only operate on the assumption that what it states must not be taken in a literal sense, and therefore it is essentially metaphorical. Whereas a topographic map sheet is 'like' the ground surface it represents, a world map is not 'like' the surface of the sphere. In this respect there can be no simple correspondence between the representation and any description of reality. The problem so posed cannot be solved by the representation. The use and usefulness of the map depend heavily on the willingness of the map user to accept the conventional arrangement, and to compensate for its deficiencies.

The representation of three-dimensional surfaces

The representation of the three-dimensional surface of the Earth is one of the most challenging and interesting cartographic problems. It remains challenging because there is no theoretically perfect solution, and because the more detailed the representation of the relief, the more likely it is to interfere with other map information. The actual variation in elevation of any part of the Earth's surface can be measured from a datum. Although specific elevations at points can be given by numerical values, only a limited number can be selected for inclusion. Lines of particular elevations, normally at fixed intervals (contours) can also be added, and these make possible the interpretation of slope to some extent. These lines of elevation can be reinforced or even replaced by showing elevation zones in different colours. Shading can be used to suggest variation in slope. But even if all are combined it is only possible to approximate the total variation in relief.

Not surprisingly, the problems of relief representation have been closely studied by cartographers, and in some cases by map users. The classic account of cartographic relief representation is by Imhof (1965). The interpretation of relief from the map symbols is an important aspect of many map-using activities.

Three-dimensional phenomena

The same basic problem of representing three dimensions in two also arises in any representation of three-dimensional phenomena which may form part of the map content, in the sense that other phenomena may extend continuously over the Earth's surface, or can be treated as continuous for illustrative purposes. Although the relief of the Earth's surface is the most obvious example, being virtually static and visible, the same representational devices can be applied to phenomena which are invisible, intangible or even subject to rapid change and variation. Temperature and precipitation are obvious examples, even though strictly speaking precipitation is not a 'continuous' phenomenon. In both cases specific values are measured and recorded at particular points, either continuously or at fixed time intervals, and given a sufficient density 'contour' lines can be interpolated.

The same cartographic practice can be applied to other phenomena which are not continuously distributed at all, but are treated as such for representational purposes. Maps of population density are an obvious example. The variation in quantity over areas is regarded as the 'third dimension', giving rise to a 'statistical surface' (see Robinson *et al.*, 1978). In these cases what the map actually states is far removed from what the map user is expected to understand. Where mean statistical values are attributed to areas, and these are then treated as point values through which isolines can be interpolated, the map descends to absurdity, as the lack of correlation with the actual characteristics of the phenomenon are apparent.

An important representational problem, therefore, is that the same cartographic device is applied in quite different ways. One isoline may well use the same line symbol as another. But the difference in quality and type of information on which the isolines are based is not made apparent to the map user – they are all equally 'convincing'. More discretion in the use of signs by the cartographer could make a significant difference to the understanding of the map user, and the conclusions drawn from the map representation. The point is worth consideration, as many 'scientific' maps are the least 'scientific' in their use of graphic symbols. A set of agreed rules or conventions, which would require that only measured lines would be shown by continuous line symbols, and that all interpolated lines would be shown by interrupted line symbols, would go some way to making the use of the sign system more rigorous.

Three-dimensional objects and features

In normal map use the viewpoint of the map user is assumed to be perpendicular to the Earth's surface at every point, i.e. orthogonal (Fig. 31). Thus individual objects or features on the Earth's surface appear in plan only. Although this retains correct angles, and therefore shapes, such a viewpoint is often uninformative about three-dimensional objects. Therefore it is not surprising that there has been a long history of the use of other types of perspective for representing three-dimensional objects in both maps and pictures.

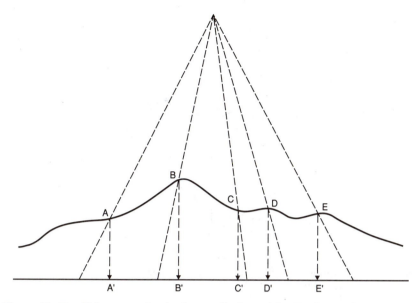

Figure 31. Parallel perspective (orthogonal) view of the Earth's surface projected onto the map plane compared with the 'bird's eye' view

Perspective and projection

Because the term 'projection' is used in drawing to describe the method by which a given perspective view is formed, it is necessary to distinguish it clearly from the use of the term as applied to the geometrical structure of the map. The *Shorter Oxford English Dictionary* makes this distinction, giving as an additional description of projection, 'The representation of any spherical surface on the flat, i.e. of the whole or any part of the Earth, more fully called map projection.' Perspective is defined as 'The art of delineating solid objects upon a plane surface so as to produce the same impression of relative position and magnitude, or of distance, as the actual objects do when viewed from a particular point.' Although not stated, this essentially defines perspective in terms of the central linear perspective now generally accepted as 'realistic'.

The use of perspective to produce the illusion of three-dimensional objects in space as perceived from a single viewpoint is only one of the possibilities of pictorial representation, and in terms of human history, relatively recent and limited. In dealing with the geometries of representational art, Hagen (1986) provides a detailed analysis of all the methods of projection onto a two-dimensional surface (types of perspective). They are classified into four main groups, depending on whether the projection lines converge to a point or are parallel, and whether the plane of the objects and the plane of representation are parallel or oblique. For example, in the representation of a cube, one face only may be shown correctly, either in plan or profile. Oblique linear (central) perspective can show three sides, but only by distorting

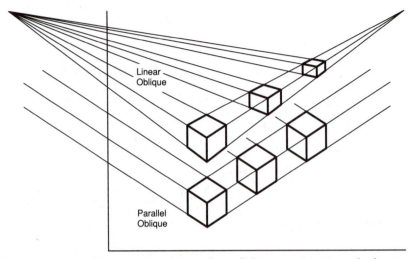

Figure 32. Central linear perspective and parallel perspective views, both at an oblique angle

both angles and distances; oblique parallel projection can also show three sides, but angular change is constant and parallel lines remain parallel (Fig. 32). In both cases an elevated viewpoint is normally used, although this oblique angle can vary considerably. Although central linear perspective gives the appearance of reality from a single viewpoint, the parallel oblique perspective actually retains more of the characteristics of the object. A high oblique parallel perspective was used extensively in both China and Japan, even though the landscape paintings were generally synthetic compositions rather than attempts to portray a real scene.

Although the first description of central linear perspective is usually accredited to Alberti in the fifteenth century, White (1957) shows that although other Italian artists had already begun to construct pictures on these principles, they had also composed pictures using more than one viewpoint. In this respect, the adoption of a single linear perspective came into existence relatively recently, and is still by no means universal.

Before this development in fifteenth-century Europe, there had been no clear distinction between map and picture. For example, the combination of two viewpoints was common in Egyptian wall paintings. Pictures were not conceived as being modelled on a particular scene, primarily because Egyptian art was not concerned with a 'factual' world. The representation of the human figure was normally a combination of frontal and profile views (Fig. 33) despite the apparent anatomical contradictions. They painted what they knew, rather than what they saw. Significantly, minor figures in the composition were often drawn 'correctly'. The same device can also be observed in Amerindian art. Mendelowitz (1970: 14) gives an example of an engraved gorget in which the same combination of profile and frontal views occurs in the representation of the human figure.

Figure 33. Egyptian figure

The well-known Egyptian drawing of a garden pool and surrounding trees (see James, 1985: Fig. 57) combines both plan and profile views. The trees, shown in profile, are placed at right angles to the edges of the pool, as were the sides of large buildings, such as castles, in many early European maps. The modern map frequently employs the same device, although usually the 'vertical' objects, such as trees or buildings, are given the same orientation. It is evident that the map user is quite content with such a use of two viewpoints, accepting the orthogonally viewed plan as the basis of the map, but allowing individual objects to be viewed in profile, or even in linear perspective.

Because the plan representation of buildings is so often uninformative, for a long period many maps continued to present buildings, individually or collectively, either frontally or in oblique perspective. This is most evident in the 'picture maps' of towns in the sixteenth and seventeenth centuries produced in many parts of Europe, where the habits of the landscape painters are quite apparent in their treatment of towns and buildings. In this respect they separated the representation of the surface, in plan, from the representation of features on it, in linear or oblique perspective.

Oblique views and cartographic representation

Although many pictorial symbols used on maps appear to be single cases of profiles or outlines, they can also have the effect of giving the visual appearance or sensation of an oblique view. For a long period, both large- and small-scale maps of topography made use of such profiles to indicate the presence of three-dimensional objects, such as

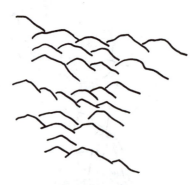

Figure 34. Hills in profile, giving the impression of an oblique view

hills and trees, by repeating a single iconic symbol. Indeed the outline profile of the hill, either as a standard symbol or with variations, was for long the preferred way of showing hills and mountains. As such, the representation treated hills as individual objects located on the map plane, rather than as part of a continuous surface. It can be argued that this is closer to the way that most people 'see' hills than a representation of a continuous surface relief. As McGrath (1965: 88) says, there is an 'important visual association between the profile symbol and the conventional view of the Earth-dweller.'

The common practice in post-Renaissance Europe was to place these hill outlines horizontally and in a single orientation. Because these symbolic hill profiles do not change in size with distance, as they would in the central linear perspective of a picture, and as one outline may interrupt or obscure another, they give the impression of a parallel oblique view (Fig. 34).

The same device is used in many modern maps of cities, where the buildings are shown in elevation. In order to be able to present, at least in part, the sides of the buildings, an oblique parallel perspective is used, as this maintains a constant scale, and parallels remain parallel.

CHAPTER 6

The map as a correct description

Given that the primary use of the map is as a source of information that can be interpreted by the user, whether or not the map can be held to be correct is of central importance to both map user and map maker. Justifiably or not, the map is normally regarded as 'correct' (i.e. the user should be confident in the locational facts presented about the phenomena described) unless the user comes to believe that this is not the case. The commonest complaint about any general map is that something on the map does not agree with what is known or believed to be the case by the user, or that something expected by the user is not there. This is turns leads to doubts about the whole of the map and its correctness. Whereas at one time it was obvious that many items on maps were fictitious or imaginary, one result of the power of modern surveying is that correct maps of the topography are known to be possible, and therefore correctness in locational information is held to be a normal property of the map. This tends to be extended by inference to other kinds of phenomena, even though the locational basis for such information may be of quite a different order. It is also extended to small-scale maps, such as general maps of continents or the whole of the world, even though in fact they are compiled from different and often disparate sources, and the results of projection mean that any simple notion that areas, distances and directions can all be 'correct' is quite untenable.

It is part of the propensity of the map that similar levels of presentation by symbols at least suggest similar degrees of correctness in the source material. A 'contour' at 5000 metres in the Himalayas looks just as convincing as a contour at 1000 metres in Europe, and this assumption is made unless the map positively states that this is not the case. With such maps, it is impossible for the user to make any visual comparison between map and subject, and therefore the information has to be taken on trust. This is, of course, one reason why maps are essentially rhetorical – a point which will be subsequently discussed.

The accurate map

The transition from maps which were only approximately correct over small areas to maps based on precise survey over large areas was achieved by the long and continuous development in surveying methods, and knowledge of the shape and size of the Earth, from the sixteenth century onwards. Not surprisingly, the claim that the surveyor could make a new and 'correct' map was frequently advanced as the chief reason for financing a new survey and publication. Historically, the acquisition of funds for new surveys was frequently dependent on the argument that an accurate, or a more accurate, map was necessary, and would be valuable to society. The prolonged but abortive effort to establish a cadastral survey of England and Wales in the early nineteenth century was based largely on the premise that accurate large-scale plans were needed by the Tithe Commissioners, and that once produced they would be of value to many users. As Kain (1975: 82) points out, Dawson, as the protagonist for the tithe survey, insisted that 'the tithe maps should portray both the tithe district and tithe area boundaries with absolute accuracy.'

Even today, many new world atlases will claim some superiority in correctness of information or treatment of projection as part of their publicity. In some cases, as with road atlases, a level of correctness is essential to the user, as they are only of value if they do indicate the current state of affairs.

Inevitably, this conception of correctness is most clearly and easily expressed in relation to 'accuracy', defined by the plan position of precisely measured points and elevations, by which locations on the map are compared very carefully with those produced by a high-quality survey at appropriate scale. But the concept of accuracy is also present in representational art. In the drawing school, the student is expected to progress to the stage where any delineation is accurate in the sense that it corresponds both to the visible form of the subject and the perspective used. Developing this facility, with pencil, pen and brush, has always been regarded as a necessary foundation for more advanced work, enabling the artist to concentrate on personal interpretation and expression.

The combination of good drawing and the desire to give an impression of 'accuracy' is another trap for both cartographer and map user. Intricately detailed representations, however inappropriate to both scale and generalization, may give the superficial impression that the map has been produced with great attention to detail, and can therefore be regarded as correct. It is an obsession often demonstrated by nineteenth-century engraved maps, many of which employ unnecessarily fine lines and intricate forms, which actually add nothing to the informational value of the map. They give an air of detailed elaboration which is intended to influence the user.

Accuracy, scale and generalization

If the basic definition of the term accuracy is accepted – 'apply care to, or execute with care' (*Shorter Oxford English Dictionary*) – then the term could be used with regard to the correctness of all kinds of maps. But the subsidiary definition, 'conforming to a standard, or to truth; precise, correct,' nowadays has an inescapable association with measurement and measurement standards. In this sense all small-scale maps can be regarded as inaccurate, because they have to depart from any simple representation of the actual forms of the phenomena. In addition, the projection of a three-dimensional surface onto a plane must involve distortion, which in turn means that a map cannot be regarded as an accurate representation of the real world. This distortion is of course the price that is paid for the convenience of having a single continuous representation of the Earth's surface, or a large part of it. In addition, because small-scale maps are so useful, the fact that the representation of shapes has to be considerably simplified and much detail omitted is also accepted.

A map user confronted with a map of the world 'knows' that the surface is actually continuous, despite what is apparently stated by the map. Even so, it is unfortunately the case that many unsophisticated users do regard the geographical positions of features as being correctly shown on the map, thereby assuming that the presentation is correct in the superficial sense. It is largely because of this that maps can be both intentionally and unintentionally misleading.

It is perfectly reasonable to maintain that a small-scale map of any kind can be correct in relation to its scale, provided that the effects of both projection and generalization are understood by the map user. Maps are neither more nor less correct than other kinds of generalization. Reduction in scale itself is not so unfamiliar to the map user, because virtually all the photographs and pictures he has viewed will have been reductions, less than life size. In the case of the photograph, generalization is either a result of objects being omitted because they are below the resolution of the recording system, or because the detail in the photograph is below the visual resolution of the viewer. With paintings, artists are aware that effects cannot be achieved by trying to copy the scene or object in every minute detail, because the process is self-defeating. Pope (1949: 61) points out that a drawing by Turner of Edinburgh 'illustrates the principle of omission of detail for the sake of emphasis of the main contours of the successive planes.' This generalization brings about what he calls 'a transposition from the terms of nature into the terms of drawing.'

Generalization and the map user

Generalization simultaneously affects both informational content and the form in which it is expressed. Whatever practices may be adopted to make it more consistent, it is essentially subjective. A simplified line, for example, cannot be objectively equivalent to some feature in the real world, because such a line does not exist. As such, the form given to it is only one out of many possible choices, and indeed what is chosen in a particular case will also reflect the type of map and its overall purpose. Even in the use of language, the stressing of what is important and the diminution or exclusion of a mass of detail is a significant element in clear description. Although the subjective nature of general-ization is often regarded as a kind of deficiency, generalization is unavoidably important because it is the means by which the bewildering detail of the experienced world is brought into a comprehensible order. As such, it is present in all kinds of description and representation.

It is unfortunate that in current cartographic literature the term generalization is often used to mean only simplification. It is unfortunate, because in many ways line simplification is the easiest aspect to understand, and also the easiest aspect for the cartographer to deal with. In practice, generalization involves five related procedures – selective omission, simplification, combination, exaggeration and displacement – and in many cases they are all applied simultaneously. Although they involve familiar cartographic tasks, they also need to be understood by the map user.

Selection and selective omission

The process of selection is often referred to as the first stage in preparing the content of a map, but arguably it operates at different levels. All maps are selective, in the sense that they deal with some things and not others. As Crick (1976: 130) says, 'no map is a total representation of reality in the sense of charting all its features. All maps are selective because there is a "point" to their construction which makes only certain kinds of phenomena significant.' Obviously a geological map 'selects' geology, but this would be of little value without some selection of topographic information as a reference base. A topographic map is directed towards the representation of the topographic features of the Earth's surface, but it will include many items that are invisible (local boundary lines) or intangible (contours). The selection of content in turn will reflect the scale and the particular characteristics of the area covered. It also depends on the judgement of relative importance from the user's point of view, and in this respect two topographc maps of the same area may differ in selected content and its classification.

The primary selection is consequent on scale and purpose. A very small-scale map, such as map example A (see Fig. 1) can include only a small proportion of the total range of phenomena of the Earth's surface. In such a map, roads, of which there are likely to be many thousands, are not the subject of attention, and therefore are entirely omitted. On the other hand, some major physical features (coastlines, large rivers) and major cultural features (large towns, international boundaries) are virtually obligatory within the purpose of the map. These categories may have concepts with very simple intensions, but their extensions (their denotata) may receive highly variable treatment.

In a medium-scale topographic map of a more specialized kind, such as map example B (see Fig. 2), the subject matter, topography, requires selection of features from the visible, permanent landmarks of the topography, especially those which form barriers to human movement. If it is a map, as opposed to a plan, then some representation of the variation in relief of the Earth's surface (i.e. the actual surface to which other features are related) must also be included. The principal features of the cultural environment are also essential (towns and roads), and therefore there are minimum selection requirements generated by the nature of the map itself.

Selection can be much more rigorous at a secondary level, for that part of the map content which has a minor and supportive role. For a special-subject map, such as map example C (see Fig. 3), some representation of the basic topography of the area is required. But what this will include will be largely a function of the geographical characteristics of that particular area. It can be said that the only absolute rule for selection will be the inclusion of major land–water divisions, and major socio-political divisions, such as international boundaries. Land–water divisions and the territories of countries are the most familiar 'units' of the Earth's surface, and their role in recognition is a reflection of their importance in selection.

Selection, designation and classification

Selection of particular phenomena is only part of the process. Selection of the concept (i.e. what the sign designates) raises questions about what aspect of the feature or phenomenon is to form the basis of the classification. This, of course, is a fundamental consideration with characterizing signs, but it often proceeds from many assumptions.

To take a familiar example, it is clear that the term 'river' is a general concept, having a simple intension and therefore an enormous extension. There are thousands of features included in the category 'river'. A large river on a large-scale map will be shown with both shorelines as boundaries between land and water. At a smaller scale, and for smaller rivers, the river channel is symbolized by a single line, although this may vary in width to suggest larger or smaller rivers, based on their two-dimensional extent. But a river may vary seasonally in width, depth and rate of flow, all of which are aspects of the particular river. A river may be contained within banks, but at other times spread over a flood plain. For some periods it may not flow at all. Except in specific cases, the degree of seasonal or annual variation is not represented. A map user looking at a topographic map, and trying to anticipate the character of the landscape, may have to make allowances for seasonal variations and other aspects which cannot be simply represented by one map symbol.

The total information that can be shown, for example by a line symbol, in representing the characteristics of any feature is limited by the nature of the sign vehicle, the symbol itself. The elaborate symbol for a major road may indicate several different aspects by use of the graphic variables: its width, class or number of carriageways, its surface type, whether or not it is fenced, and its place in a route classification. All these are possible within the capacity of a multicolour symbol. But the 'width' of a road symbol, especially at small scales, is not necessarily related to its physical width on the ground at all, but its relative importance, on either an objective or a subjective basis. In this respect the sign not only overtly designates, it appraises, and the aspect that is appraised is selective. An area symbol may describe 'natural' vegetation according to its physical constituents (e.g. woodland) or by its usage (e.g. rough pasture). Even on general maps, the mode of characterization of a symbol may select one aspect in preference to another. What is signified, for example, by the designations 'swamp' and 'marsh', and would the interpretation always be correct?

Selection also extends into the classificatory process. This is inevitable, because very often as selection changes, so does the type of classification. Although roads may appear on both large- and small-scale maps, the basis for selection will be different. Whereas on the former they are likely to be classified on the basis of their physical and operational characteristics, on the latter the emphasis will be on their functions as routes between major places. The same designation, road, will be applied to features which are classified and selected in a different way.

Denotation and omission

The next part of the generalization process, omission, is more straight-forward in principle, but more complex in its effects. Selective omission refers to the restriction in denotata of the feature or phenomenon that has been selected and designated. To use the same example, a map that includes 'rivers' will not necessarily show all the features in the area which fall within that designation, but only a certain proportion of them (Fig. 35). How this proportion is arrived at will also vary from one part of the map to another. Omission is a function of scale, geographical density and relative importance. Therefore it is uneven in its effect. Map example B (see Fig. 2) shows most, but not all of the minor motorable roads in the area. In map example A (see Fig. 1) the density of geographical features in the same class in relation to scale enforces

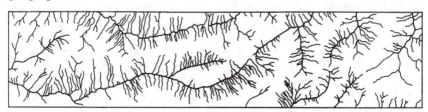

Figure 35. The need for selective omission; drainage at 1:50 000 scale reduced to 1:250 000 scale

rigorous omission. Most of the large cities in Japan have been omitted, even though they are larger than many of those shown in Manchuria. The process is dictated by the limited space of the two-dimensional field of the map, and it presumes the user's understanding. The effects are most difficult to predict on many medium-scale maps, where omission may only be slight, and limited to some areas – a fact which may not be obvious to the map user.

At large scales, omission has to operate even at the data collection stage. In a topographic map based on an original survey, some decision has to be reached in the original specification as to the smallest size of feature in a particular class which is to be recorded. This applies particularly to areas of natural vegetation and land use. If this is based on measurement, it can be systematically applied. But the difficulty with many maps is that whether or not a feature is included will depend on a judgement of relative importance. In this respect omission is appraisive, and involves a subjective element which is usually accepted as characterizing generalization.

Locational information and simplification

Although the map shows 'where things are', the map representation may have to differ considerably from the forms apparent in reality. The consequence is that both linear features and outlines presented on the map have to be made less complex than the external reality they represent. Inevitably, the effect is most pronounced at small scales, but it also operates at large and medium scales. It is also uneven in its effect, because the more irregular a feature, the more simplification will be required. A perfectly straight section of road, appearing on both a large-scale and a small-scale map, will be straight on both; only the represented length will be different. A winding road with many bends may be shown 'correctly' at the larger scale but must be considerably simplified at smaller scale. In consequence, both its length and sinuosity will be reduced (Fig. 36). Simplification of the outlines of areas has the same result. As complexity of shape is reduced, so the outline becomes less detailed, and the overall perimeter decreased.

It is in maps used directly in comparison with the landscape that the consequences are most serious for the map user. The need for the simplification of complex and irregular shapes has to be accepted and adjusted to. On a medium-scale map, such as 1 : 100 000, although the measurement of distance may be minimally affected by projection, the measurement or estimation of length along a feature will be affected by line simplification. The process applies to both indexical signs, such as contours, and characterizing signs, because it is a function of scale and two-dimensional space.

In cartographic practice, every effort should be made to ensure consistency in simplification. If the application of simplification differs between two sets of phenomena, then one may appear to be more symmetrical and regular than the other, and this will be reinforced by the apparent 'detail' of the less simplified forms. Where contour lines

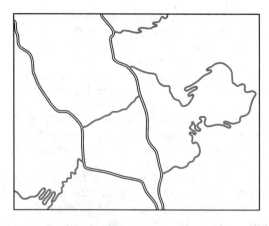
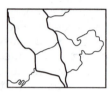

Figure 36. Simplification: the representation of roads at 1:200 000 scale and at 1 : 500 000 scale

show the shape of the surface, the drainage pattern needs to be adjusted so that it accords with the simplified contours, not treated as a separate item. The same applies to the location of settlements in relation to drainage or slopes, and the relative position of roads and physical features. Modification of the features of the physical landscape by simplification therefore has to be matched by correct positioning of the cultural features of the landscape. Like so many other cartographic operations, although they have to be carried out singly, they have to be thought of and viewed collectively.

This is also a problem with many small-scale specialized maps, in which the topographic reference base is taken or copied from an existing map quite separately from the compilation of the specialized subject matter. Frequently the topographic base is too detailed for the simple forms of the primary subject.

Combination, exaggeration and displacement

When individual small features belong to a category that is important in the map content, the effects of omission and simplification may be counteracted by combination, exaggeration and displacement. Small adjacent areas of one type may be combined into a single unit, by eliminating the small spaces in between them (Fig. 37). On the other hand, in built-up areas, the narrow corridors of the main roads may be exaggerated in width in order to retain them in a legible manner, and to show their relative importance. The same process extends into, and overlaps, symbolic exaggeration. Whereas at a large scale an individual feature, such as a building, will be shown in its correct plan dimensions, at a small scale this will be replaced by a point symbol which is proportionately much larger than the actual dimensions of the feature to scale.

The need to retain features that would be too small to be included at

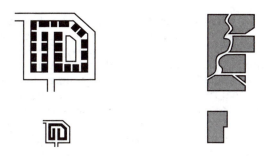

Figure 37. Combination and scale reduction

map scale, resulting in the deliberate enlargement or exaggeration of a linear feature, may lead to the slight displacement of other features which are positionally related to it. A symbol for a building that is located at the roadside has to be moved to remain in its correct relative position if the road is indicated by an enlarged symbol.

The process of exaggeration, by enlarging the apparent dimensions of features which are too small to be shown according to their true dimensions at map scale, is one of the most powerful elements in map composition. By this means, what is important in the subject matter is retained, and where necessary emphasized. The selectivity of the map is reinforced by concentrating on some things at the expense of others. Selectivity and exaggeration can make the map much more informative than an aerial photograph of the same scale, despite the apparent 'objectivity' of the latter. The selective judgements underlying the processes of generalization are not simply a graphic necessity, but are one of the map's most powerful attributes.

Generalization and the fourth dimension

It is customary to think of generalization as essentially the modification of features on the two-dimensional map plane, even though these may actually be three-dimensional. This is natural, given that the normal map is a static image, and that maps compiled from other map sources all exist as two-dimensional representations.

The consequence of this is that any feature can be shown only in one position, whether it is a linear feature, an outline or a boundary. But natural phenomena range from those which fluctuate in position over short intervals (such as the diurnal movement of many shorelines), and those which fluctuate seasonally (e.g. snowfields, seasonal lakes), to those which fluctuate irregularly over longer time periods (such as temporary watercourses and inundated areas).

At large and medium scales, lines representing both high and low water are likely to be shown in tidal regions, both of which are either mean or typical values. At smaller scales the generalization of this inter-tidal zone is normally done by using the high-water mark, (i.e. the limit of permanent land), although of course there are always exceptional states when the limit of water will not coincide with this fixed line. Long-term variations in precipitation, especially in very dry regions, mean that the 'edge' of a lake may well be in quite different positions in different years or different decades. Although the blue line often used to represent a river graphically suggests the presence of water, many seasonal rivers may be dry for a large part of every year.

In many specialized maps of phenomena which are in a constant state of change, all lines indicating specific values have to be based on mean values or cumulative totals. For example, mean annual precipitation is characterized by averaging the totals over a cycle of years to arrive at a representative figure. Mean monthly temperatures, so often shown in series to represent the annual temperature variations at a given location, have to be calculated from daily recordings. Although graphs of the actual temperature fluctuations would be more 'correct', the generalizations are necessary in order to make a mass of data comprehensible to the reader, and it is these generalizations that are used to construct the isotherms on maps of temperature. In this respect many statistical maps are themselves generalizations, even though this is rarely suggested in their description.

The difficulty for the map user is that the same symbol type, the continuous line, may be used to represent a moving edge, a fluctuating zone, or a mean value based on a series of records or measurements. The user is expected to make the correct interpretation, even though the lines, as symbols, are superficially similar.

Unintentional error and misunderstanding

Given the complexity of the generalization processes, it is hardly surprising that the treatment of the information can lead to misunderstanding on the part of the user. At the unintentional level this may simply be the result of cartographic incompetence. This incompetence may be due to the fact that the producer may be insufficiently aware of the connections between information, scale and generalization, or simply regards them as unimportant. One only has to look at the haphazardly simplified outlines of many maps used in the mass media to see the inadequacy of such maps. It is unfortunately true that large numbers of mainly illustrative maps are made or controlled by people who are unaware of cartographic principles.

Coulson (1977: 101) points out that 'A pervasive characteristic in the perception of maps (and for that matter, diagrams) is the apparent objective, factual nature of the image presented.' He cites an example of unintentional misrepresentation in which the scale of a map used to illustrate statistics was inadequate for the purpose, resulting in significant omissions, and therefore offering the reader misleading conclusions. Inadvertent misunderstanding can also occur when displacement is carried to extreme lengths. The simplification of shapes and position in the famous London underground 'map' (which is really a diagram) achieves simplicity of description by distortion of shape and position. Because it is accepted as a 'map', which implies that positions are correct, the user may be surprised to realize that two stations apparently far apart may be almost adjacent.

Maps and propaganda

The capacity of the map to mislead has also been exploited to give false impressions of geographical 'facts'. Such maps cover a wide range, from the relatively innocuous use of some maps in advertising to propaganda for powerful ideologies. Some obvious examples are found on stamps. Given the frequency of disputes over national territories and the positions of national boundaries, it is hardly surprising that postage stamps – which are widely circulated – are deliberately employed to make territorial claims which are founded on expectation or ambition, rather than the *de facto* position.

Probably the classic case of maps used as propaganda was their employment by Nazi Germany. As Whittlesey demonstrates, much of this was incorporated in the monthly journal *Zeitschrift für Geopolitik* from 1924 onwards. A host of pseudogeographical terms was coined to support the case for German expansion. Whittlesey gives an example of a map from 1934 in which Bohemia–Moravia is shown as apparently having the same elevation as the Alps, forming a huge mountainous block between Berlin and Wien, and so threatening the German state.

In contrast to this deliberate abuse of the map, the focusing of attention upon military conflict can lead to a need for a better understanding of geographical relationships, and therefore strategy, on the part of the general public. Hendrikson (1972: 23) describes the importance of journalistic maps in explaining the geographical relationships between allies and enemies in the Second World War to the American public. The famous maps by Harrison, with their oblique projections and perspective views, were intended to inform and educate the public, although they could also be regarded as propaganda for the war effort. As he says, 'Harrison's distinctive strategy was to portray the Earth from so many radically varying angles of vision that viewers would redevelop a native freshness of perception.' Much of this was necessary to counteract the undue influence of Mercator's and other cylindrical projections in general maps of the world, which used uncritically tended to produce 'a horizontal and east–west based' conception of the relative positions of the land masses.

Map quality

The idea of a 'good' map always implies that the information shown is locationally correct and complete, as well as being presented in such a way that the user finds it intelligible. It is perfectly possible for good information to be poorly presented, and for poor information to be well presented. It is usually difficult for the map user to distinguish between the quality of the information and the quality of the presentation. It would be desirable in cartographic practice to make greater efforts to reveal to the map user the quality and nature of the basic information by a more rigorous use of graphic symbols. For example, all lines which indicate mean values, and especially isolines interpolated from randomly scattered point values, could be shown by interrupted lines, leaving the continuous line to represent only those features which are constant, have definable positions and are based on measurement. It is commonly done to distinguish between permanent rivers and seasonal rivers, and it could be applied more widely to great effect.

PART THREE

The map as a communication

CHAPTER 7

Theories and models of cartographic communication

As part of the general theorizing about cartography, much attention has been directed to the relationships between the activities of map makers on the one hand and map users on the other, the map itself being regarded as part of a communication system. During the 1970s and early 1980s this led to the development of general theories of cartographic communication, all of which were essentially attempts to reveal the connections between the creation of the map as one set of processes, and obtaining information from the map as another. The theories themselves have certain elements in common, but also distinct differences.

In the first place it is necessary to consider the sources of this development. Map making, like other applied arts and sciences, does not exist in isolation, and necessarily reflects changes and developments in society as a whole. In the last 30 years or so the growth and increasing importance of communication through mass media has been recognized, and these media have largely depended on advanced technology. With the rapid development of computers, the storage, retrieval and processing of vast quantities of information takes place at ever increasing rates, and computers have become an accepted part of the fabric of the modern State. Their capacity and sophistication naturally generates comparisons with the human brain, and indeed the term 'memory' is applied to both. The increased supply of information of all sorts is regarded as an essential basis for many kinds of economic, social and political action, and the problems of obtaining, organizing, and making use of such information are of major consequence in a huge range of human activities. Not surprisingly, consideration has also been given to the efficiency with which information is used, and the degree to which it is successfully employed, whether this is a statistical analysis of a business operation, or the effect on sales of a televised advertisement.

Along with this, the various sciences making up psychology have also developed rapidly, and an improved understanding of how people behave and react to their environment is accepted as important. The deliberate influencing of human behaviour for political, social and economic reasons is actively pursued, and this in turn is influenced by assumptions about how people react to external stimuli. Earlier general theories of the S–R type (that a certain stimulus leads to a predictable response) have largely been abandoned as the complexity of perception has been demonstrated. Current research in perceptual psychology and physiology continues to yield greater knowledge of how the central nervous system functions, and indeed the first part of this book is an attempt to apply some of this knowledge to questions of visual perception and map use.

The parallels between these large fields of activity are clear. In a communication 'system' there is an input of information which is transmitted by one means or another to a receiver, which in turn produces an output. A human being receives 'information' through the senses. This is duly processed in some way, which leads to a change in state or activity. A computer is supplied with an input of data, which is duly processed according to a program, through which an output can be obtained. As a map is a source of information that can be perceived by a user, then presumably it should be possible to analyse the input, transmission and reception of map information as a 'system'.

Theories of cartographic communication

Of the many theories and ideas that have been put forward, the most influential are probably those of Koláčný, Ratajsky, and Robinson and Petchenik, with developments and additions by Morrison. Although the analogy with 'communication' had been noted by Board (1967), Koláčný (1969) was the first to put forward the suggestion that making and using maps should be treated as a whole, and that 'cartographers' should be concerned with the use of maps as well as their construction. In common with many others, Koláčný made no attempt to define 'cartographer', and it must be assumed that he referred to map makers in the widest possible sense. One of the unfortunate consequences of the general vagueness of terminology is that these general theories refer to 'cartographic' communication. But whereas the broad use of the term 'cartographic' must include all aspects of map making, including data collection, the term as employed in the theoretical models often suggests that it is being used in its narrower sense, i.e. the graphical design and construction of maps.

Many of Koláčný's strictures on the attitudes and practices of 'cartographers' have been widely quoted, but the origins and objectives of his research must be appreciated if they are to be kept in context. Koláčný had been concerned for many years with the functions of maps within the general educational system in Czechoslovakia, and had concluded that

The contents of newspapers, radio and television broadcasts have been confronted with the knowledge imparted by primary and secondary school education. Both young people and adults were found unable to understand the geographical content of current information disseminated day by day by the mass means of communication, unless they could draw upon the requisite information from independent work with maps (Koláčný, 1969: 49).

The evidence for this is presented by his further paper (Koláčný, 1971), which attempts to show the disparity between news items involving a knowledge of location (geographical concepts) and information available in school atlases. His conclusion was that neither the maps nor the map users' abilities were sufficiently developed to provide a proper basis for understanding world affairs, and that therefore: 'The task of modern cartography is on the one hand to supply the schools and public with such systems of maps which would optimally help the teaching of geography in schools and in self-education and on the other hand to study the methods of map utilisation and improve them' (Koláčný, 1971: 222).

It is not altogether clear as to why Koláčný believed that these deficiencies in the employment of maps were due to the attitudes of 'cartographers'. The use of maps within formal education would seem to raise problems not within the power of cartographers to solve. Whether maps are made, and to what extent they are available, is rarely in the hands of 'cartographers'. If he meant that the pupils found the maps difficult to work with, then the criticism would have some substance, but he does not provide any evidence for this. Consequently, when he states that 'the work of the map user is largely determined by the cartographer's product. There prevails the tacit assumption that the user will simply learn to work with any map which the cartographer makes. In other words, the map user is expected to submit, more or less, to the cartographer's conditions' (Koláčný, 1969: 47), it is difficult to know whether he is referring to examples of bad or unsuitable maps, or how he arrived at this 'tacit assumption'. Nor is it clear whether the 'conditions' are the product of a cartographer's arbitrary decisions, or the consequence of map structure which the users might not have understood.

Map information

Koláčný's diagram showing the stages by which cartographic information is communicated has long served as a basic model. He takes the starting point as the cartogapher's selective observation of reality. This 'multi-dimensional intellectual model of reality' is transformed by the cartographer's mind into cartographic information, objectified and expressed by map symbols. This is therefore available to perception, and can be increased in availability by 'map multiplication'. The map in turn produces an informative effect on the map user, transforming the user's opinion about reality, so that the map user creates in his mind a 'multi-dimensional model of reality' and experiences this reality. Acting on the strength of this cartographic information the map user either transforms this into practical activity, or 'processes it into an idea which he realises at once or later on' (Koláčný, 1969: 49).

Koláčný emphasizes the importance of cartographic information as opposed to 'map content', by which he seems to mean the total gain in knowledge by the map user. He indicates the difference between the cartographer's 'reality' and the user's 'reality', and also the need for these to overlap. At this stage it was clear that he was concentrating on a general statement of the processes involved in making and using maps, and he does not refer to transmission or coding. As the titles of his papers show, he was mainly concerned with demonstrating the importance of cartographic information, and with emphasizing the value of maps in a proper understanding of world affairs.

The development of communication theories

Koláčný's appeal to cartographers to be more concerned with the provision and employment of maps acted as a powerful stimulus for further discussion and development. The idea that cartography should be extended to include all aspects of making and using maps was fully expressed by Ratajsky (1973: 217–227), who maintained that cartography should be regarded as 'part of the informatics or the new large science of communication.' Based mainly on the ideas of Board (1967) and Koláčný, his model of 'cartographical transmission' adopts the concepts and terminology of theories of information transmission, summarized in his statement that 'The messages are transmitted from the cartographer (K) to the receiver (O) by means of a channel of transmission, i.e. by means of a map (M)' (see Fig. 38).

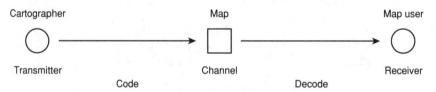

Figure 38. The basic concept of some theories of cartographic communication

This is then elaborated in a complicated description, in which Ratajsky attempted to include all the factors in the sequence between reality (R) as the source of information, and a 'reflection of reality' created by the receiver (O). Between reality (R) and the map (M) he refers to 'message emission' (the relationship between the cartographer and the transmission means, i.e. the map); and 'message perceiving' (comprising technical processes that lead to the construction of the map). He also states that 'a loss of information occurs determining the efficiency of the message emission.'

Between the map and the map user he identifies 'indirect emitting' (consisting of errors, imperfections and technical flaws), and 'indirect perceiving' (depending upon the 'receiver's disposition and upon the external conditions of map viewing'). These also lead to a loss of information. The final stage takes place by 'the identification of symbols, by the recognition of messages included in them, and by a comparison of these messages' so that the ultimate result of the correctness of imagination (R^1) depends on the 'ability of recalling' and the 'ability of imagination'. Finally he summarizes all this by saying that the upper relations, R–K and O–R, are concerned with map contents, while the bottom ones K–M–O are concerned with map form and the correctness of the map language. He then goes on to a general description of theoretical and applied cartography, map theory, map knowledge, and so on, which is more or less a catalogue of anything to do with maps, and which does not add anything to the basic proposition of information transmission.

Interpretations of the communication model

The next major contribution to this development was by Robinson and Petchenik (1975). Whilst accepting the general notion that maps are concerned with communication, and that therefore the importance of this should be recognized by cartographers, they were quick to qualify some of the more simplistic tenets of the communication analogy, especially the relevance of information theory and the concept of 'noise'. In reviewing various expressions of the general communications 'network', including the stages of encoding, decoding and transmission of signals, they give various interpretations of this analogy with the functions of maps. Already several discrepancies become evident. According to Ratajsky, the map is the 'channel' of transmission, and therefore the cartographer is the 'transmitter' and the map user the 'receiver'. One interpretation of the analogy according to Robinson and Petchenik (1975: 9) is that 'The world and the cartographer constitute the source, the map is the coded "message", the signal is made up of the "light waves" which make the message visible, the channel is space, and the decoder and recipient are the receiver-destination'. It is clear that in this interpretation, the map is not the 'channel', and transmission is seen primarily in terms of visual perception, i.e. the light reflected from the map and reaching the eye. But this is not analogous to 'transmission' in the communication system sense, in which signals are encoded into a form suitable for transmission, and are therefore imperceptible to a human observer. Although Robinson and Petchenik (1975: 11) point out that 'when the cartographic communication process is analysed in detail it is evident that it departs significantly from the general model', it is still by no means clear as to how this general model should be applied. Ratajsky, of course, does not explain how 'transmission' actually operates, but this also applies to other aspects of his theoretical model.

Robinson and Petchenik contribute to the discussion along three important lines. First, they give a detailed account of why information theory (as a means of measuring the efficiency of information transmission) is a false analogy as an explanation of communication through maps. They conclude quite firmly that 'Because the communication of the "information" on a map is in no way like that of a coded sequential message consisting of signals, the measurement of the 'information content' of a map and the amount transmitted by a cartographer to a map percipient cannot be effectively obtained with the techniques of Information Theory' (Robinson and Petchenik, 1975: 13). Second, they distinguish between map readers, map users, and map percipients. Whether or not these definitions are accepted, they emphasize the different levels of map-using operations. (It is of interest to note that the three types so described are roughly equivalent to map examples A, B and C; see Figs 1–3). And third, they develop a different model of cartographic communication.

To do this they use a Venn diagram in which the outer rectangle represents 'the total conception of the geographical milieu held by mankind' (Robinson and Petchenik, 1975). This is divided into correct and incorrect segments (Sc and Se). Area A represents part of this milieu in the sense of the 'cartographer's conception', and area B is the conception of the geographical milieu held by a percipient. The overlap between the two indicates a degree of knowledge or understanding common to both. This overlap seems to refer to geographical knowledge rather than map structure. A small rectangle (M) denotes a particular map, regarded as a coded message, which contains three subdivisions: M1 consists of information in the map which is already familiar to the percipient. M2 is the information in the map which is new to the percipient and therefore an addition to his understanding; M3 is the fraction not comprehended, which symbolizes 'the discrepancy between input and output in the communication system'. Finally, the area U, which lies beyond the rectangle representing the cartographer's conception, stands for 'an increase in spatial understanding by the percipient that may occur because of his viewing the map but which was neither intended nor symbolized in any way by the cartographer' (Robinson and Petchenik, 1975: 11). This of course indicates 'interpretation' by the map user, in which statements made by the map stimulate associations, correlations or conclusions derived from other knowledge held by the percipient. If this can take place – and it clearly does – then how does it fit into a conception of cartographic communication in which 'messages' are transmitted to a receiver? In this case the response depends upon the map user and cannot be deliberately intended by the map maker.

The notion advanced by Ratajsky, that the study of cartographic communication leads to the development of cartography as a 'scientific' subject, capable of sustaining its own theoretical development, is reinforced by Morrison (1976: 84–97). Maintaining that cartography should be seen as 'an information or communication science', he accepts Ratajsky's structural model of cartographic transmission, and identifies 'at least two realms of activity in a communication process: the initiation of the communication and reception of the communicated message.' He makes the point that the initiator of the map can also be the receiver of the communicated message, but of course in Ratajsky's model the initiator of the map is the cartographer, and the 'overlap' is between reality (the source of the information) and the map user's imagination of reality. In a more simple diagram than Ratajsky's, Morrison (1976: Fig. 2), distinguishes between reality (I); the cartographer's cognitive realm and the cartographer's concept of the map (II); the map and the cartographic language (III); map interpretation, consisting of information read from the map and the map reader's cognitive realm (IV); leading back to reality. The initiator of the map is stated to be the cartographer, and the linear model of information flow is maintained.

In developing his theory in greater detail, Morrison (1976: 90) asserts that 'cartography begins with the cartographer's desire to communicate a portion of his cognitive realm to someone else.' The first stage is selection, in which the cartographer 'makes a conscious selection of part of his cognitive realm which he wishes to communicate to a map reader.' This is followed by a process of generalization (including classification and simplification). The map is prepared as a communication channel by the use of cartographic language, which must be known to both the cartographer and the map reader. At the same time, the cognitive realms of both should have some information in common, and it is this factor which permits communication. This component is defined as 'base data'. If the map does not possess the base data as assumed by the cartographer, then inefficient communication results. A similar breakdown occurs if the cartographer fails to include sufficient base data. Both are described as the fault of the cartographer.

The idea that maps can only function if some geographical information is already held by the map user leaves a major question unanswered. What happens if a map user looks at a map of totally unfamiliar terrain? Whatever 'base data' the cartographer includes, there may be no common element so far as geographical knowledge is concerned. Yet it is perfectly possible to use such a map, and of course the user can do so by understanding the map's locational structure and the system of signs which indicate it.

Further pursuit of the communication analogy leads to familiar problems. Morrison observes that if the map reader is not 'fluent in the language' then communication breaks down. In addition, he accepts that the cartographer 'has no control over induction' and that map interpretation takes place 'within the cognitive realm of the map reader.' The effect of this on the idea of messages being sent to a receiver is not considered.

Morrison (1976) then puts forward the proposal (first given in Morrison, 1974) that a table could be constructed in which each map symbol could be specified in terms of the physiological and psychological responses of the map user, depending on the map reader's ability to perform 'detection, discrimination, recognition and estimation in a spatial framework.' The proposition, suggesting that map symbols can be specified in some grammar regardless of context, and that a standardized map user can represent individual human beings, is a prime example of the behaviourist approach.

Finally, he distinguishes once again between map reading and map interpretation. On the one hand there is the statement that 'It remains an open question as to how much the actions resulting from map reading are a part of the cartographer's concern.' But in a further paragraph there is the statement that 'the reading of a map and the mental processes involved in interpreting the map are all legitimate cartographic concerns' (Morrison, 1976: 97). In all attemps to model cartography on a simplistic communication theory of output equalling input, interpretation remains a dilemma.

Map evaluation

Although Board (1967) was one of the originators of the concept of cartographic communication, his later version of the model is largely based on Morrison. In a later study (Board, 1977), however, he is mainly concerned with the analysis of map reading and interpretation, stressing the importance of processes such as identification, verbalization, visualization and interpretation. This underlines the positive role of the map user, and the activities of both anticipation and verification. In relation to the problem of achieving communication between the map maker and the map user, he points out that failure in communication may be due to either 'the cartographer's poor map-making ability, or the map user's inadequate skill at map reading' (Board, 1977: 55). Some of these ideas are further examined in a later paper (Board, 1978a), which is directed towards what he calls 'map evaluation'. In this he attempts to analyse map-reading tasks and develop 'reasonably objective tests of the effectiveness of different forms of map.' Although under the general heading of cartographer communication, these studies are therefore devoted primarily to the elucidation of the map user's role in real map-using operations. This search for positive information is in sharp contrast with the general vagueness of communication theories.

Cartographic models and communication theory

The use of the term communication has become widespread in the modern world, especially in the sense of mass communication and the notion that something called information is distributed rapidly and constantly. Indeed, communication skills are now being stressed in a great deal of education and training in Britain. Significantly, skill in reading is seen to be just as important as skill in writing and speaking. Yet the term communication, like many others which have expanded in use and application, has reached the point where it has ceased to have any specific meaning. In itself, it is an excellent example of the way in which language, far from operating on the basis of a single and agreed sign/referent for every term, can be debased to the point where it is virtually meaningless.

Sless (1986) points out that in its original usage 'to communicate' meant to make common or to share. In later centuries it added a further meaning by reference to means of communication, such as roads, canals and railways, these being the physical structures along which things were passed and therefore shared. This was the source of the idea of communication as 'conveying', initially in the literal and physical sense. By extension this has become confused with transmission, which also refers to the movement of things along lines, or in channels, in a form which is imperceptible or intangible. Therefore communication can be understood either as the idea of sharing, of holding things in common, or as transmission, a one-way process, directed along specific lines or routes.

Both these elements were present in the original ideas of cartographic communication, in the sense that the physical model of information transmission was combined with the notion that the cartographer shared or made common geographical information through the map. Because such models were essentially framed in language, and described by diagrams of components and movements, the physical or mechanistic model of transmission was dominant. The genesis and development of these ideas has been described in several papers, in particular by Board (1983), which also includes a useful bibliography of all the main contributions. Its various interpretations have been discussed and argued over in many fields, but there was little reflection of this in its introduction into cartography as an instance of this general development. Indeed the lack of any serious analysis of the fundamental concepts was quite remarkable. Fisher (1978) shows that there are many possible interpretations of terms such as meaning, message, information, feedback and process, and that 'information' can be used in at least three different ways. The fact that such a simplistic model of making and using maps could be so rapidly adopted says more about the general social and academic climate of the time than cartography as such.

Writing about psychology and sociology, Sless (1986: 12) observes that these two disciplines were 'greatly preoccupied with their scientific status.' He goes on to point out that

> It was clear that the natural sciences and their highly successful technological offspring were taken as models for these younger subjects. Even when the theories and findings of these admirers of so-called hard science were vague and inconclusive there was a tendency to present the subject using terminology and concepts that had the appearance of science and precision. The emerging studies of the psychology and sociology of communication were no exception and the favoured model used to account for human communication was taken from electronic communication engineering.

Despite the numerous variations on this theme, Sless (1986: 12) points out that all these models 'break up communication into three domains: all the activities or processes which go toward the making of messages, the domain of the message itself, and the domain of the receiver of the message.' These therefore reduce human communication to 'a series of manageable areas each of which could be isolated as a component and studied separately or in relation to the other components.' The ultimate model, therefore, is a closed system, in which all the elements are fully controlled, resulting in perfect or error-free 'communication'. It is a description which could be applied directly to a great deal of theorizing about communication as applied to cartography.

In a philosophical sense these accounts of communication as a system are essentially positivist. Miller (1987) maintains that this reflects the belief in the existence of general rules which operate regardless of the particular case. Yet when these scientific principles are applied to the explanation of human conduct, they inevitably raise the question of whether or not the human 'sciences' are fundamentally different to the natural sciences. Indeed, whether the study of any form of human behaviour can be classified as a 'science' is a source of disagreement. This is not a problem limited to cartography and map use; it has been discussed in relation to history, economics, psychology, anthropology, sociology and geography. Dreyfuss (1986: 11) makes the point succinctly: 'Inevitable difficulties arise for the human sciences not from the definition of man but in so far as the current sciences of human beings imitate physical science in seeking a theory that precisely predicts events in the everyday world using context-free features abstracted from that world.' As Heelan (1983: 16) points out, this perspective is dependent on the assumption that 'the methodology of the positive sciences is in principle capable of answering all meaningful questions.'

Geography itself, with its long-standing division into 'human' and 'physical', also lay in the middle of this argument. Indeed the re-examination of geographical theory, especially in the 1960s, was one instance of the triumph of the natural science model, as human and social geographers rapidly adopted the apparatus of the 'pure' sciences. In this respect it was not surprising that many geographers sought to bring the map, and therefore cartography, into the same ambit. As Pickles (1985) pointed out, 'Once the criteria and standards of what constitutes knowledge and progress in the natural sciences had been adopted by the social sciences, all preceding systems of thought and theories were automatically regarded as pre-scientific; belonging to the realm of social philosophy, metaphysics and unscientific speculation.' It was in this sense that Harvey (1969: 374) complained that there had been an 'almost total lack of consideration for the logical properties of the map as a form of communication,' maintaining that 'The map, is, in short, a stimulus which is designed to elicit a certain response.'

In contradiction to this, Capella (1972: 232) suggests that

> Law-like explanations of communication behaviour, for example, treat man as a stimulus–response system; given the stimulus–response law and the appropriate conditions, the response must follow. Now, while a large segment of behaviour is habitual, unconscious and unintentional and, therefore, well suited to law-like explanations, there exists a second segment which is purposive and choice-oriented and, therefore, poorly suited to law-like explanation.

Given that theories of cartographic communication have to accept and find a place both for the intentions and responses of the map user, and the need for interpretation, these factors cannot be fitted within a closed and predictable system.

Several other elements characterize these descriptions of cartographic communication. There are frequent references to the 'language' of maps, and the function of signs in cartographic representation. Both linguistics and semiotics have been explored as possible sources of theoretical structures that would explain the nature of the map and the way it functions. The analysis is essentially reductionist, through which the whole is subdivided into smaller and smaller units. According to Rose (1986), the reductionist methodology breaks down complex systems into their constituent parts, but he argues that although this has been immensely profitable in the natural sciences for the past 300 years, it is incapable of predicting what may happen when such units are combined in complex interactions. Petchenik (1977: 117) makes the same point in relation to cartography: 'The assumptions underlying the work of experimental psychologists for the past 40 years or so have been essentially reductionist rather than wholistic, and behavioural rather than dealing with mental processes.'

Major advances in science have always depended on imaginative thinking, not just analytical research. Yet a reductionist approach has been pursued even to the point of looking for the cartographic equivalent of morphemes and other units characteristic of certain forms of linguistic study. This happened at the very time that many philosophers were insisting that communication could not possibly operate in this way, and that the analysis of one sign system (the map) by the use of another sign system (language) poses an insuperable obstacle to any claim to objectivity.

In a simple sense there would seem to be a parallel between author–text–reader and cartographer–map–map user. Both involve the creation of an object which is 'read' and interpreted. Both involve skill in creation and interpretation. Leaving aside for the moment the question of whether a linear form is truly analogous to a non-linear form of representation, there are fundamental questions about the nature of a text, and its relationship to both its author or creator, and its reader or user. In a detailed examination of the basic premises of semiotics, Sless (1986: 34) maintains that 'At the heart of communication lies a paradox: the text both joins and separates authors and readers.' And he goes on,

> If we now look at the author's position . . . we can see that any statement he makes about readers must be a construction – imagined – and any notion of the readers which he uses either implicitly or explicitly to guide the construction of his message must also be imagined. This needs to be understood as a general condition of communication, not as a communication failure that can be put right by replacing imagination with knowledge – as if there was some method by which authors can be brought closer to their readers (Sless, 1986: 35).

Yet 'replacing imagination with knowledge' is a crucial element if the cartographer's apparently inadequate skill is to be replaced by a set of rules based on scientific analysis.

To some extent, the deficiencies of these early models of communication through maps have been recognized. Yet references, usually unexplained, to cartography as 'communication' still persist. Except in rare cases, what a map offers is a possibility, not a message. The problem facing the cartographer is essentially that of representation. It is only by solving the representational problems that the map, as an ordered structure, can contribute to any communication process.

CHAPTER 8

Communication, map use and cartographic objectives

Scientific theory and cartographic communication

The desire to elevate cartography to a scientific 'discipline' in its own right, complete with its own theories, is in some respects a typically academic pursuit. The danger is that a crust of pseudo-scientific terms may conceal a whole set of misconceptions and simplifications. The development of any comprehensive theory needs to be based on evidence, and at present there is a distinct lack of evidence about how people actually use maps, and what problems they encounter, which is a key element in any communication process. At the same time, it must be recognized that although we cannot satisfactorily explain how good maps are made, or how some map users are able to employ them successfully, this does not mean that successful map use does not occur. Anyone who has watched a first-class orienteer navigate across unfamiliar terrain to pin-point accuracy and at a high speed does not need any further evidence that maps can be and are used successfully. It is also clear that such a performance requires a combination of a suitable map and a proficient user.

Any scientific theory should have certain properties. To select only a few, it should be universal, in the sense that it can account for all known instances; it should be economical, avoiding the introduction of unnecessary terms; it should be based on evidence, derived either from observation or experiment; it should be capable of development, to allow for refinement and future changes; and it should define its terms. In general, it must be said that most of the communication theories demonstrate few, if any, of these properties. The most striking deficiencies are the failure to scrutinize the analogy of information transmission to test its validity and to see whether it can serve as a basis for both explanation and development; the omission of several important factors; and the absence of any attempt to provide examples from actual maps to show how communication is supposed to work. Although some authors clearly make assumptions about how maps are made, there is little indication of the range of activities and circumstances which can lead to map production, nor the wide range of activities which can be grouped under 'map use'. Unless a basic communication theory can be shown to account for what is already known, it is unlikely to inspire any confidence in its capacity for future development.

Communication models

The concept of a communication system as a linear transmission process is essentially the view generally described as 'mechanistic'. In the words of Fisher (1978: 100), 'the mechanistic ideal of the concept of 'process' is little more than a temporal sequence of events within a closed system.' He continues (1978: 101),

> It takes little imagination to realise mechanism's profound appeal to the practising scientist. If prediction is a desirable goal of scientific enquiry (and you will recall that prediction is the ultimate goal of some scientists), then conceptualising the observed system as closed and therefore deterministic is a natural theoretical perspective for the conducting of empirical research.

Fisher points out that describing any model of communication seems to lead naturally to some sort of diagram, with arrows linking the various parts; that (1978: 107) 'the perspective of communication held by the vast majority of people, including laymen and scholars alike, includes a strong dose of mechanism,' and that (1978: 133) 'The mechanistic perspective of human communication emphasises the physical elements of communication, the transmission and reception of messages flowing in a conveyor-belt fashion among source/receivers.' This mechanistic view overlaps what Fisher describes in general as the psychological perspective on communication. In this, the sensory reception of stimuli leads to greater emphasis on the way in which the organism responds, or processes the incoming signals. But although the objective of an S–R explanation is again prediction, the complexity of factors involved in the response is recognized. Fisher (1978: 162) observes that in this approach 'the internal mechanisms of perception and information processing are the prime forces of communication.' In this respect the map is seen as a means of achieving certain responses on the part of the receiver (map user), and therefore there is a need to carry out psychological tests of the human perceiver to find out how he responds. In considering the theoretical analysis of communication, Fisher concludes (1978: 163) that 'the bulk of writing on or about communication is probably a blend of mechanistic and psychological elements, probably with a greater emphasis on the psychological.'

The influence of these attitudes can be seen in the communication theories applied to maps. They are just as questionable in this application as they are, taken by themselves, in any other. Although the development of theories about cartographic communication is just a small part of the theorizing about communication in general, there is surprisingly little discussion of such theories, or of their suitability for an analysis of making and using maps.

The analogy of information transmission

It can be argued that it is perfectly proper to advance theories or models without realizing any immediate practical advantages from them. Indeed, scientific research and development would be impossible on any other basis. Nevertheless, having advanced a theory *a priori*, the next stage surely is some attempt to test it against known facts, or to devise an experiment to see if it 'works'. As a starting point, therefore, it is of interest to test the type of communication theory advanced by Ratajsky against what is commonly known about making and using maps.

The basis of the analogy is that something called 'information' is transmitted from a source through a channel to a receiver. As many authors have pointed out, the concept was originally applied to the problems of radio and telephone communication systems, and the primary objective was to reduce the loss of 'information' between the sender and the receiver. It must be noted immediately that in this sense 'information' is a purely technical matter, and has no connection with meaning; and that information theory was developed to provide a system of measurement to quantify the efficiency of such a transmission system.

The characteristics of such systems have to be carefully examined. The source or starting point may be a series of sound waves which are audible, e.g. human speech or an orchestral performance. The object of transmission, as the term implies, is to send these sounds over a distance, and subsequently make them audible to a listener not within hearing distance of the source. Whether the transmission is by broadcast radio waves, or along telephone 'lines', the original sounds must be encoded into signals which can be transmitted. In this form they are no longer perceptible to human hearing, and therefore must be decoded and transformed once again into sound waves perceptible to the listener.

Because of the need to transform one type of energy into another (required by the channel) and because of external factors (such as atmospheric conditions), the output at the receiver may depart significantly from the original input. Unwanted signals, which are not part of the information, are described as 'noise'. The perfect system should produce a facsimile of the original source, but as every hi-fi enthusiast knows, this is technically impossible, despite the advanced technology devoted to perfect reproduction. In such a system, the meaning of the information is immaterial. If a person speaks gibberish into a telephone, and it is clearly audible to a listener, then the communication system is working perfectly well. It is, however, questionable whether any 'communication' has taken place. It must also be noted that after the technical 'receiver' has performed its functions, its output still has to be perceived by a human 'perceiver', and it is at this point that comprehension takes place, or a response made.

In this sense, the expression 'communication of information' has two quite distinct meanings. In the colloquial or general sense it means to 'share knowledge about'; in the technical sense it means to transmit and receive a distinguishable signal. Eco (1976: 8) makes a distinction between communication and signification. He points out that we 'define a communication process as the passage of a signal (not necessarily a sign) from a source (through a transmitter, along a channel) to a destination. In this sense we have no signfication, but we do have the passage of some information.' He goes on to say (1976: 20) that 'the proper objects of a theory of information are not signs but rather units of transmission which can be computed quantitatively irrespective of their possible meanings, and which therefore must be properly called "signals" and not "signs".'

The same distinction is made by Black (1975: 151), who deals with the term 'information' in a discussion of the nature of representation. He comments that

> The typical situation to which the mathematical theory applies is one in which some determinate stock of possible 'messages' which may be conceived of as alternative characters in an 'alphabet' (letters, digits or pulses of energy) to which no meaning is necessarily attached, are encoded into 'signals' for transmission along a 'communication channel' and ultimate reception, decoding and accurate reproduction of the original 'message'.

He adds in a footnote that 'It is ironic that experts in information theory have repeatedly protested, apparently without success, about the misleading consequences of identifying what is called 'information' in the technical theory with the meaning of that word in ordinary language.'

Information transmission and feedback

In many cases the model of information transmission introduces the term 'feedback'. This process is commonly applied in electronic control systems, and refers to the modification of the signal which is the output. Its indiscriminate use in discussions about communication fails to make clear that the function of feedback is to modify the signal before final output. It can hardly be applied to the reactions of a map user to a particular map. This has to exist before the user can operate with it, and any critical views or comments can only be acted upon by making changes after the map has been produced.

Information transmission and the map user

Is such a system of transmission applicable in any sense to the use of maps? If the object of transmission is to convey the original message to distant receivers, then the only parallel is the multiplication of the map into many copies, which are physically transported to the users. Apart from technical flaws, such as misregistered colours, the significant fact is that the printed image is not converted into some other form in order to be transmitted, but is a virtually perfect duplicate of the original. If the map is transmitted by television, then of course it is affected by coding into signals and subsequent decoding by the television receiver, and is subject to the limitations of colour and resolution of the television channel. But this form of communication is readily understood to be a technical matter.

The view that a map user can be regarded as a 'receiver' in the sense of reproducing a facsimile of the original is quite without foundation. If there is one point on which all authorities are united – philosophers, psychologists and physiologists – it is that the human percipient is not a passive receiver of messages. As Bellin (1971: 99) puts it, the perceiver 'is neither passive in his reasoning, nor is he a passive recipient of stimulation from the physical world,' and 'perception is not a passive encounter with external stimulation.' Neisser (1976: 13) states that 'Perception itself depends upon the skill and experience of the perceiver – on what he already knows in advance.' Crick (1976: 138) comments that 'we know that human beings are not passive recorders; our perception is selective and organisational.' Rescher (1973: 2) observes that 'in all our attempts to portray reality, the output is a function not only of the "external" inputs alone, but of the mind's transformational mechanisms as well.' Bertin (1978: 123) makes the same point: 'one should move into action in order to get all the information held in a diagram or a map, and not remain a passive receiver.'

It is conceivable that in some cases a map user might attempt to memorize an entire map, or more probably part of a map, thereby 'receiving' all the information. Even in such an event, a very simple map could only be remembered imperfectly for any length of time, and in most types of map use no such attempt at complete storage takes place or is considered.

The alternative analogy, as illustrated by Robinson and Petchenik (1975) and others, avoids any reference to transmission in this sense, and applies the same general ideal to the direct perception of the map by the map user. In this case the signals are light waves and these are 'decoded' in the mind of the map user. Although no 'transmitter' is referred to, the signals must be sent if they are to be received and decoded, and in this respect the map maker and the map are still regarded as the 'active' components in the process, as the cartographer originates and conveys messages to a receiver or percipient.

But light waves reflected from the surface of the map and passing through 'space' can hardly be described as having the channel characteristics of the transmission theory. Optically there are no limitations on the range or amount of information which can be transmitted. What can be perceived is controlled by the limited capacity of the perceiver, that is, the map user.

A radio receiver tuned to a given wavelength cannot reject or accept certain 'messages', or choose to attend or not attend. If a signal is transmitted there will be a predetermined response and the term 'signal' is quite correct in this context. But it has been explained already that map symbols and signs are not signals, and what the user responds to is affected by the perceiver's knowledge, attention and ability. Therefore if the idea of direct perception is applied to map communication, the active roles of transmitter and receiver have to be reversed. The map itself is passive. Communication takes place only when a map user actively directs attention to it.

It is in this respect that a proper appreciation of how signs function is important in understanding how communication is achieved through a map. Peirce maintains that the function of a sign is triadic, and that the sign mediates between the actual phenomena and the perceiver. This can only take place through a process of interpretation. As Greenlee (1973: 104) expresses it 'the kind of determination appropriate to sign activity must be distinguished from bare causal relation. Mechanical reactions and reflex responses, whether conditioned or innate, are not, in the appropriate sense defined by Peirce's theory, interpretative of an action or a stimulus.'

It can be pointed out that the linear or mechanistic view of communication – messages being transmitted from a sender to a receiver – has also been widely discussed as a theory relevant to other fields of study. Although McQuail (1975: 14) is primarily concerned with the social sciences, his comments on the value of this type of communication theory in regard to human behaviour are highly appropriate to the use of maps, and are worth quoting at length. Having observed that, according to information transmission theory,

> Any communication act involves a sequence of events which takes the basic form of a decision to transmit meaning, the formulation of the intended message into a language or code, and an act of transmission and of reception by someone else.

He goes on:

> At the same time we should bear in mind that this view of communication may be highly misleading, however useful for exposition. Misleading not only because most communication events are much more complex than this model suggests, but also because of a bias written into this particular formulation. For one thing, it implies an underlying rationality and purposefulness about communication, an intention of achieving certain ends and hence the relevance of a criterion of communication efficiency. Second, it proposes a linear view of communication which is inconsistent with what frequently happens. Third, and correlatively, it suggests that communication always begins with a sender, and that it is the intention of the communicator which defines the meaning of a communication event.

The first of the points made by McQuail is a valid comment on the communication system as envisaged by Ratajsky and Morrison. The second deals with the linear model, which Petchenik (1977), Robinson and Petchenik (1975), and Guelke (1977) strongly refute. The third point is particularly important in relation to the initiation and preparation of a map, which Ratajsky, amongst others, sees very definitely as an act of deliberate formulation by an individual cartographer. But whether the activities and responses of map users can be predetermined by map makers is a central question in the whole idea of a communication system as applied to maps.

McQuail goes on to say (1975: 17) that

> The term 'receiver' in our basic communication model is in many ways misleading. It has a very limiting connotation, implying as it does a passive role, one defined primarily in terms of reaction and response. It also appears as a role defined largely in terms of the communicator, and hence lacking in autonomy, as if it could not exist apart from a sender.

He continues:

> The receiver is, more often than not, also an initiator, both in the sense of originating messages in return, and in the sense of initiating processes of interpretation, with some degree of autonomy. The receiver 'uses' and 'acts on' the information available to him. In another important sense he is an initiator, in that he has chosen to attend to some sources and messages rather than others.

The value of the simple communication model as a basis for the examination of relationships between map makers, maps and map users is largely questionable. Although Castner (1979b: 36) points out that 'there are so many variables in the picture that it may be impossible, given our present understanding, to incorporate all of them clearly and succinctly into a model of cartographic communication,' he nevertheless takes an optimistic view of the usefulness of this model as a general basis for discussion. But the physiological and psychological evidence for regarding visual perception as a directed, discriminating, interpretative process is so strong that it is highly unlikely that any such model can be employed as a firm foundation for further studies. Indeed, the constant need to qualify and interpret, and to introduce new concepts and terms, illustrates how difficult it is already to employ any linear model satisfactorily. In other respects the model is also directly misleading, for it largely conceals the important contribution of the map user's active role.

The initiation of the map

Despite the imperfect foundations of the transmission analogy, can it provide some sort of outline which can illuminate the sequence of operations involved in making and using maps? Board (1978b: 48) invites 'those interested in cartographic communication to relate theory and practice.' This can be taken up by considering at the outset the question of how maps originate, and therefore how the communication system is apparently set in motion.

According to Koláčný the process starts with the cartographer's selective observation of reality; according to Ratajsky, the cartographer becomes acquainted with reality; according to Morrison (1976: 89), the starting point is 'the cartographer's desire to communicate a portion of his cognitive realm to someone else.' Despite the variations in expression, the impression is given of a personal or individual cartographer who wishes to make a map for some purpose known to him, with the object of communicating this information to one or more users. It is not explicitly stated, but in none of these accounts is the term 'cartographer' either qualified or defined. It must also be assumed, for lack of any statement to the contrary, that the cartographer is acting on his own initiative, as the map user is not regarded as playing any role in the initiation of the map. The viewpoint is commonly expressed by geographers writing about cartography. It can be exemplified by Dobson's description (1977: 39) of this approach as taking the view that 'The reader's systemic input, in turn, in extracting information from the map is an attempt to realise the cartographer's objectives in presenting the display.'

It is fairly easy to see how this viewpoint arises. If a scientific map author is writing a text dealing with spatial information, he is quite likely to include maps, in which the spatial relationships of the phenomena are described, or the author's theories about them represented. In these circumstances the map author can be seen to be the 'cartographer', and the map is the product of the author's ideas and intentions. As Board (1977: 47) points out, 'Geographers frequently initiate maps to illustrate their arguments about distributions and relationships.' The intention is that the reader, who is also the map user, will be invited to consider the maps, the content and design of which have been selected and differentiated according to the wishes of the author. In this context it is understandable that the map information could be regarded as consisting of 'messages' sent by the author to the reader. The term 'cartographer' may be applied to the map author, who initiates and controls the map content, or it may be applied to the cartographer draftsman (cartographer) who frequently produces the map, and may take some part in its actual design, as well as its preparation. The use of the term 'cartographer' remains confused, because relatively few scientific map authors would describe themselves as cartographers, even though they may on occasions practise cartography. For example, geologists make no such claim. The point may seem trivial at this stage, but its consequences are considerable.

This type of map making and map initiation is perfectly valid in its own right, but can it serve, even by implication, as a model for other types of map?

If other examples of map making are considered, then it becomes clear that in most cases the map is not initiated by a 'cartographer' at all. Contrary to Koláčný's assertion (1969: 47) that 'the user will simply learn to work with any map which the cartographer makes,' in many instances the initiator of the map is also the map user, or some body or institution representing his interests and needs. To take some obvious examples, virtually all national topographic map series have been and are produced to satisfy the requirements of government departments, military agencies, local authorities, and a variety of 'pressure groups', ranging from engineers to tax collectors. The fact that most of them may understand little or nothing of how the map should be made, or even what sort of map would best serve their interests, does not alter the basic condition that map production requires resources, and ultimately the decision to devote public money to map production does not lie within the control of 'cartographers', however defined. Nor is this a recent development. Throughout the nineteenth century and up to the present, the Ordnance Survey of Britain has been subjected to numerous enquiries and assessments. Although 'cartographers' have been called upon to give evidence, the numerous committees of enquiry have been composed almost entirely of individuals representing map users, or professional experts with map requirements. The content of the map (its 'information') is normally their primary concern.

To take another example from modern practice, let us assume that a government minister in a developing country has decided that some undeveloped region should be mapped, primarily to provide basic information for economic and social planning, and possibly military defence. He invites tenders from private surveying companies to carry out the mapping operation. He, or more likely a representative committee, states the objectives, the purposes which the map is to serve, and the survey companies explain (and cost) what will be required to reach these objectives. In many cases the initial scheme will be modified or reduced, not because of any lack of understanding of what the intended users need, but because basic mapping of this type may be more costly than the particular community can support. In such a case, how does the statement 'The cartographer makes a conscious selection of part of his cognitive realm' (Morrison, 1976: 90) apply? Cognition implies knowledge, but the reason for the mapping in the first place is the absence of any real 'knowledge' of the area. In such cases the relative positions of the map user and the map maker are clearly defined by the Royal Institution of Chartered Surveyors (1980): 'Clients should select the scale, accuracies and content of the mapping to satisfy their requirements, leaving contractors to select the survey methods which will achieve these specifications both reliably and economically.'

Other examples are not difficult to find. Some of the most important maps of all, from a user's point of view, are the group of special-purpose maps dealing with navigation. They range from very small-scale navigational charts (both nautical and aeronautical) to very large-scale charts of harbours and airfields. In Britain the first of the 'official' coastal chart series was commissioned by the Admiralty in the eighteenth century and carried out by private marine surveyors. At the beginning of the nineteenth century an official Hydrographic Department was established, specifically to provide naval ships with coastal charts. The requirements were stated by the Admiralty on behalf of the chart users. The cartographers, i.e. the marine surveyors and chart engravers, were brought in to carry out the work.

Even in those circumstances where the interests and needs of the map users cannot be determined exactly, strenuous efforts are made to anticipate them correctly. In the case of a private company, making and publishing maps for profit, the sale of any map will largely reflect the ability of the company to anticipate correctly what the public will buy. In this case the initiation of the map may depend on the judgement of a 'cartographer'. But the objective is the satisfaction of a known or presumed user requirement, and many private companies go to great lengths to determine as fully as possible how best to achieve this. Although Koláčný (1969: 47) states that 'the creative work of the cartographer should be based on the fullest regard to the needs, interests and subjective conditions of the map user' as though this was a new idea, most professional cartographers in private companies would be surprised to learn that they were engaged in doing anything else.

Although topographic maps, nautical and aeronautical charts, road maps, and general reference atlases probably provide most of the maps which are actually used for some purpose, there are still other types of map initiation. In many scientific fields, particularly the Earth sciences, the importance of maps as information stores is realized. To take another example of map initiation, geological map series are usually produced by a geological mapping agency, frequently attached to a geological institution at national level. The maps produced serve to advance the scientific study of geology, and also provide fundamental information for many other activities, and therefore a wide variety of uses. They are usually constructed, in the first instance, by geological surveyors making observations on the ground or interpreting aerial photographs. The primary requirement is information for the study of geology, and not the promotion of some 'cartographer's' personal interests.

Personal motivation does play a role in the initiation of maps. There are some individuals who see maps not only as scientific documents, but as elaborate and sophisticated products of human skill and ingenuity, with an aesthetic appeal in their own right. Just as architects design 'beautiful' but functional buildings, so many cartographers believe that maps can be both functional and works of art. Considering the efforts made by virtually all human societies to express themselves by constructing artifacts which are both functional and pleasing, it is not really surprising that maps can also be regarded in this light. It raises an interesting question, however, of whether any work of art is intended by its creator to communicate, or simply 'express'. The signs and symbols used by human beings are of course employed for both.

In modern times the activities of Bradford Washburn (1960, 1978) in initiating the beautiful maps of Mt McKinley and the Grand Canyon reflect the enthusiasm of one man for sophisticated topographic maps. Although personal initiative provided the impetus, both maps are valuable additions to the total information about those features, as well as being composed to help visitors to those areas.

In passing, it may be noted that the aesthetic properties of maps pose considerable problems for any 'scientific' communication model. In most cases the question is simply avoided. Yet any discussion with map users about many types of map rapidly raises matters which can only be described as aesthetic. Whether or not people 'like' certain maps is not simply a matter of information flow. And if a theory is to be comprehensive, then presumably it must allow for any factor which affects the user.

As the above examples demonstrate, the starting point for the initiation of a map does not arise from some single, uniform activity, but comes about in various ways in response to a determined need, which may be realized by a map maker or a map user. Broadly speaking, the types of initiation may be categorized as follows:

1. from a decision by intending map users, or their representatives;
2. from a decision by map makers, in anticipation of a need;
3. from a decision by a map author, as a consequence of her own need for explanation;
4. from a decision by a scientific body, to contribute to the total information in a specialist field;
5. from a decision by an individual to express herself through making a map.

In considering how maps originate, it is also important to remember that there are large numbers of maps which might be made, and possibly could be made, if sufficient resources were devoted to them. But in the long run it is the map user, the consumer, who pays, and the control of resources needed for map production is rarely in the hand of the 'cartographer'. The particular case of the map generated by the wishes of the map author is the exception, not the rule. In addition, many maps are not produced, not through lack of interest in the needs of map users, but because the users themselves are not prepared to support the cost.

Initiation and map content

Given that at some point the decision is reached to make a map, what is the next stage? Although all maps can be said to contain 'information', some conclusion has to be reached concerning what the map is about, and therefore its content. Koláčný distinguishes between 'cartographic information' and 'map content', the latter being the 'sum of graphical elements'. This of course raises another difficulty in the communication theory. What a cartographer puts into a map, by one means or another (and assuming that he has the necessary information), is a series of locational statements about particular phenomena. What the map user extracts will go beyond this (the actual content) to a variety of interpretations. Despite Morrison's statement (1978: 102) that 'The process of map reading then is, in one sense, the reverse of function gi, the creation of the map', the point can be illustrated by a very simple 'map' (Fig. 39).

The map content shows that two features belonging to the category 'river' drain into another feature belonging to the category 'lake', in

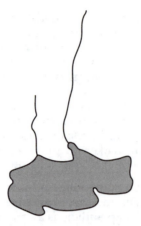

Figure 39. Individual interpretations

particular geographical locations. This content is equally visible to a set of map users. The first one says, 'Ah! one river is shorter than the other.' The second one says, 'Ah! one river is longer than the other.' The third says, 'Strange, I did not think that lake was there at all.' The fourth says, 'Interesting. Two parallel rivers so close together – probably due to geological faults.'

Now it cannot reasonably be argued that the cartographer consciously intended these 'messages'. Indeed, he could not, because the locations of the features are a matter of geographical fact and are not dependent on anything so vague as his 'conception of reality'. Presumably what Koláčný meant by content referred to the factual statements made by the map, whereas what he called information refers to the particular gain in knowledge made by the individual map user. The latter is of course a matter of the user's interpretation. What the user extracts is different to what the map maker 'puts in'. The point is covered by what Robinson and Petchenik (1975) refer to as a 'kind of unplanned increment.' Although they limit this to geographically sophisticated percipients, any conclusion or inference reached by a map user is fundamentally of the same nature.

Leaving aside for the moment the nature of what is communicated, the determination of the map content requires three things. First, the general character of the map and the purposes it is to serve have to be decided; what Robinson and Petchenik refer to as the cartographer's conception of the map. How ever the map originates, this must be at least broadly defined at some point. Second, the map has to represent a specific area. And third, a decision has to be made about its scale. The map cannot proceed at all until these decisions are made, even if only provisionally.

The omission of any reference to scale is one of the most curious characteristics of most of the theories of cartographic communication. It is not included in Ratajsky's model of cartographical transmission (Ratajsky, 1973: Fig. 1) nor in his model of cartography (1973: Fig. 3). It is not included in Morrison's 'Outline of the science of cartography' (Morrison, 1976: Fig. 2). It is not included in Muehrcke's diagram of the cartographic processing system (Muehrcke, 1972). Yet in the relationship between map and 'reality', and in most cases between map and map user, scale is the most important factor. It is scale – the ratio between what the map describes and reality – which makes maps useful (as miniaturized representations), and at the same time introduces many important cartographic problems. It is scale (in relation to area mapped) which provides the metrical problems of projection transformation; which largely determines how much omission has to take place; which underlies the whole of generalization; which limits maps to some particular purposes and excludes others; and which is probably the most difficult aspect for the user to comprehend. To devise cartographic theories without reference to scale is like trying to write about history without a calendar. Generally speaking, the decision about scale is the most fundamental decision about any map.

Given that the area, the general subject matter, and the appropriate scale have been decided, the next step is the detailed specification of the map content. This is often referred to as the 'planning stage' in map making. Presumably it is this content which is the 'input' in simple communication theory, and according to Ratajsky it seems to be something selected by the cartographer. Once again, the lack of any specific examples makes it difficult to appreciate the intended meaning. If a basic topographic map is taken as an example, how is the content 'selected'. Obviously the general subject matter is topography, and this is evidently the concept of the map, but the specification has to proceed to the particular. Once again, this is not a simple and straightforward matter. Once the decision 'topography' has been made, then to a large extent much of the content follows automatically. Neither cartographers nor surveyors 'select' rivers, buildings or roads: they are included because they are there. Their ordering, classification and symbolization, and their relative importance will vary from one map to another, but to a large extent the map content may be predetermined by the subject and scale.

The same applies to a special-subject map, such as a geological map. If this is a basic map, the normal objective is a complete and systematic description of geological formations. The criteria of how they should be distinguished and classified depend on the scientific principles of geology. It is not left to a 'cartographer' to decide what to include, and they are not dependent on a cartographer's conception of anything. The same applies to a coastal nautical chart, used for inshore navigation. Once the purpose and scale have been agreed, then a whole range of specific features have to be included (depth of water, navigational aids and hazards, etc.). Indeed it is often the inability to be selective about particular map content which poses difficult cartographic problems.

Map content and sources of information

Once the content has been planned, the next problem lies in obtaining the information. In some of the more simple expressions of communication, information is referred to as though it was readily available and waiting to be used. This of course raises once again the interpretation of the scope and meaning of cartography. If it is a synonym for 'map making' then it has to include the provision of information. If it refers to the processes of graphic construction and representation, then it can hardly account for all the issues which affect the relationship between the map and the map user.

According to Ratajsky (1973: 218), 'The source of information (R) includes direct information perceived by an observer, in this case by the cartographer (K).' Is 'perceived' intended to include measurement? Does this description imply that the 'cartographer' is also responsible for data acquisition? Koláčný (1969: 48) states that the cartographer 'either directly observes the geographical medium or he studies it on a map.' But despite the number of maps that can be generated, wholly or in part, by derivation, or composed from secondary sources, the basic information about the Earth's surface on which all maps are founded can only be obtained through topographic and hydrographic surveys. If 'observes the geographical medium' is really intended to account for all the problems involved, and the whole of surveying, photogrammetry, and remote sensing, then it is not very illuminating.

It is of course quite common to interpret cartography as subsuming all phases of map production, and indeed anything to do with maps. If this view is taken, then in a comprehensive theory of cartographic communication the importance of data collection needs rather more emphasis. If, as Koláčný suggests, more attention should be paid to the needs of the map user, then it is not difficult to find one of the sources of difficulty. One major problem lies in the fact that prospective map users (or the bodies representing them) either cannot or will not provide for the cost of obtaining the required information. The process of effective map communication breaks down, not because of any lack of regard for the needs of the user, but because of the intending user's lack of regard for the vast amount of work needed to obtain the information.

The second factor which the communication theories fail to reveal is that the information needed to make the map has to have certain characteristics, and satisfy certain conditions. The map has a dual structure, involving both location and description. Mappable information about the topography of the Earth's surface can only be obtained by measurement. Neither derived maps nor special-subject maps are possible without this basic reference material. Even those 'maps' referred to by Ratajsky and others, based on other 'surfaces', would be meaningless unless they could be compared with real maps showing the facts of actual location on or in relation to the Earth's surface.

If the study of cartographic communication is intended to illuminate the problems which arise in the satisfaction of map users, then the need for, and characteristics of, adequate information must be included. How many professional cartographers have been confronted with a request for a map, only to find that the resources available for obtaining the information were quite inadequate for the purpose? The consequence, very often, is that an inferior map is produced, which in turn fails to satisfy the user's expectations.

The need to distinguish between factors in map communication which are due to lack of cartographic knowledge or understanding, and conditions which are imposed on map makers, is essential. If maps are to be made more efficient – a plea voiced by many advocates of communication theories – then the provision of adequate information which satisfies the conditions of map structure also has to be taken into account. The existence of inadequate maps does not necessarily prove anything about cartographic theory, any more than the existence of vast amounts of literary trash proves anything about literature. It says something about readers as well as writers. Information does not 'flow' from something called reality to the cartographer. It has to be sought, and in many cases the cost and difficulty of obtaining it will largely determine whether or not a prospective map user will actually be provided with a satisfactory map.

The cost of information

So far as the interrelationship between the practical tasks of the map maker and the needs of the map user are concerned, two quite different situations exist. As has been mentioned previously, if a map author is using a map as an illustration to a text, then the purpose and content of the map are controlled by the map author. In many cases the total amount of 'information' is very limited, and is likely to be developed from the author's research material and data. But if the need for a map is identified by the map user, or anticipated by a cartographer, the question of obtaining suitable data immediately arises. Can the requirement be satisfied, wholly or in part, from existing information, for example by modifying an existing map? Will it involve the provision of new information? The difference can be critical. Does the source of information exist, and is it available? If new information is needed can it be obtained within the available budget? At this stage the original plan for the map may be modified or even abandoned, on the grounds that carrying out the initial plan would be too expensive, or too time consuming.

Similar considerations apply if a cartographer commissions a specialist to provide information for a particular map, as might happen, for example, in a national atlas. The specialist might well point out that a suitable source of information is not available, that no such information exists, or that the collection of such information would be prohibitive.

At this preparatory stage proposals for maps are devised, tested against information sources, and costed. Many prospective map users and map authors find to their surprise that the apparently simple maps they propose turn out to be very expensive, or that the scale which is technically or financially feasible is inadequate for the purpose. Whether or not the map user's requirement will be satisfied, and to what degree, may be affected by the practical considerations of time and cost in relation to the number and importance of the potential users. Even when a map is a necessary accompaniment to some research, it is not entirely unusual to find that the resources devoted to the map are less than adequate. In military mapping, where the map maker is normally working directly for the consumer, a particular map may be deemed necessary even if it is very expensive to produce. The critical point is whether the user is willing, or able, to pay.

The direct concern of prospective map users in such cases is mainly with area, scale and content. Although the representatives of the users may be quite definite about the content in general, they are unlikely to be well informed about either the problems of obtaining the information, or of composing the map. In many cases, the form of the map – its graphic presentation – is constrained by external factors, and this in turn is a direct function of the information required. The cost of production affects all products which have to be manufactured. Engineers could no doubt design the 'perfect' motor car, but those actually manufactured are a compromise between what is possible and what the user can afford. For the commercial production of a world atlas, for example, the users' interests are initially defined in terms of what the prospective users can be expected to pay. Thereafter the informational content and form are arrived at in terms of what can be attempted realistically at a given price level.

The map and reality

Most of the communication theories refer at some point to 'reality', or a cartographer's conception of reality, as the source or starting point for cartographic information. In practical terms, such references have to be accepted very broadly. However, the question of the relationship between map and reality is an interesting one, and worth considering.

One important question is whether a map is the consequence of a cartographer's conception of reality, or the source of it. According to Ratajsky, Morrison and others, the map arises through a cartographer's selective response to the real world. References to a cartographer's 'cognitive realm' do not make this any clearer. For example, one item in my cognitive realm is the belief that there is a continent called Africa. I know its general location on the Earth's surface, and can visualize it (or rather a small-scale map of it), even if only fleetingly. Yet despite the fact that I have travelled over parts of it, my knowledge of Africa as a particular portion of the Earth's surface is fundamentally map derived. But if my cognition is based on maps, how can it be regarded as preceding their establishment?

The presumption about a cartographer's personal conceptions as a source of maps is inimical to understanding how maps are actually made. The point is clearly made by Salichtchev (1973: 109): 'The making of many maps is not based upon direct observation (investigation) of reality but upon the mobilisation by the cartographer of available sources.' Petchenik (1977: 123) shrewdly observes that 'the spatial knowledge that is acquired in the course of ordinary life is multi-sensory in nature.' In this example, a personal conception of Africa is an amalgam of facts, impressions, sensations and errors. Its correspondence with the 'reality' of Africa would be difficult to determine. As a source for making maps for people to use, it would be highly unsatisfactory! Like all conceptions (as opposed to concepts), it is unique to the individual. The great advantage of the map, and the one imposed by its structure, is that it is not a consequence of a cartographer's personal conception of anything, but should be a statement of fact based on shared concepts. It can extend beyond the limitations of any perceived 'reality', or any conceptions that any cartographer can personally possess.

The question of whether there is any objective 'reality', equally accessible to all human beings, is of course disputed by philosophers. Russell (1914) points out that all knowledge is acquired either directly by perception of the external world, or indirectly by reportage. Although the description of the topography of an area, as represented on a large-scale map, can be verified by direct comparison with the terrain, the major part of our geographical knowledge cannot be verified by direct comparison with directly observed reality. If the sources of information are erroneous (like the continental outlines of medieval European maps) then the maps will be at fault. But any improvement in the representation (and therefore our knowledge) of reality comes about, not through developing personal conceptions, but by the patient accumulation of facts.

A one-to-one correspondence between reality and perceived reality is clearly impossible. From a cartographic point of view, the corres- pondence between a map and the real world has quite specific limitations, inherent in its structure. For a human being, the real world is experienced in four dimensions. The map has only two. It can fully indicate position in two dimensions, and it can partially represent the third, although the interpretation of three-dimensional form, whether of the Earth's surface or some other phenomena, often poses difficult problems for the map user. As to the fourth dimension, the map, once produced, is a static image. This again is a perennial problem for both map makers and map users. The user operates with the map in the present, but the information on it is dated to some time in the past. If the enquiry is directed to phenomena or events which change very slowly, or if information about a past state is what is wanted, this may not be a source of difficulty. But for very many map uses, the discrepancy between present conditions and what the map shows is a major hindrance. Indeed, one of the main influences on modern map and chart production is the desire to improve this by faster production and more rapid revision. One of the most important factors behind the drive towards digital mapping is the recognition of the need to provide a more rapid response in supplying maps to users. This again underlines the importance of the map users' requirements as a factor in map production.

The problem of how any impression of the external world is

represented in a person's mind is extremely complex. Even the idea that anything corresponding to 'reality' can be established from the fragmentary and limited evidence of a map is open to debate. What representations are made in the mind of the external world is a crucial question in cognition, and at present there is no evidence that this is near to a solution.

The role of the cartographer

The point has been made already that communication theories generally seem to use the term 'cartographer' to refer to all the people concerned in making a map, although whether this includes data collection is by no means clear. In many cases it is at least implied that this activity is performed by, or controlled by, a single person. The roles of map user, map author, and cartographer have been mentioned in relation to the initiation of the map, and must be further distinguished here. Although the initiation of the map, the formulation of its general concept or plan, the specification of content, and the collection of necessary data are all essential stages in the preparation of the map, there still remains the construction of the map in its final form. This must include the design and specification of the cartographic symbols and its technical production.

For some types of map it is possible, although unlikely, for all this to be carried out by one person. For most maps it would by physically impossible, and in the majority of cases the production of a map requires the cooperation of a number of people, sometimes a large number. If the term 'cartographer' is restricted to its more limited connotation (and the one most commonly used in practice) as referring to the persons responsible for devising and constructing the graphic image of the map, then it is possible to distinguish between cartographers and the many other specialists who contribute to map production.

Some examples may serve to illustrate the point. To take an instance already cited of a new topographic base map, the initiation will arise from a body representing the users, who may engage a surveying company (or their own surveying department) to obtain the information for the specified content. After the information is obtained, by field survey and photogrammetry, it is passed to the cartographers to construct the map in its final form, normally in such a way that it can be reproduced in print. The map design – that is, the use of the cartographic 'language' – is likely to be initiated by the cartographers, in discussion with the surveyors and submitted to the client for approval. The final form of the map, including the exact specification of symbols and colours, will result from discussions between the three parties.

To take another example, a research geologist may wish to produce a geological map of a given area. He will initiate the map, formulate the general concept, and be responsible for the geological information. If a suitable topographic base map is not available he may commission a surveying organization to provide the necessary base. Alternatively, he may be able to use an existing topographic base, if a suitable one can be found. Cartographers will then be involved in designing and executing the map, probably by making a trial proof to begin with, which the geologist may comment on and modify. Generally speaking, the geological content of the map will be the geologist's responsibility, whereas the cartographic presentation of the map (its visual efficiency) will be the cartographer's responsibility. Because it is impossible to separate content, design and technical factors of production, the whole map will result from a dialogue between all the parties concerned in its production.

The cooperation of three sets of specialists is by no means unusual for a map of this type. Geologists, surveyors and cartographers will all 'initiate' at some point. But it is highly unlikely that the cartographers who devise the symbols and employ the cartographic 'language' will have a full understanding of all the geological concepts embodied in the map. It is equally unlikely that the geological map authors would regard themselves as competent in using the cartographic 'language', or able to solve the problems of map design.

When these examples of quite common map-making operations are considered in relation to the statements apparently made in communication theory, several difficulties of interpretation emerge. If a map is a product of a cartographer's 'cognitive realm', to which of the parties does this apply? Is it the geologist who conceives the map, or is it the cartographer who conceives the map form? It may be argued that in a general theory the term 'cartographer' is simply intended to cover both these aspects. But if several people are involved, how many cognitive realms are concerned, and what is the level of communication between them? For someone with an adequate understanding of map making, the oversimplification might be acceptable. But would anyone with a slight knowledge of map making (such as the average map user) develop a true understanding of the 'input' into a cartographic communication system on the basis of such a model? A good basic model would lead naturally to the discussion of more complex details, but the multiple nature of contributions to the creation of a map is not made apparent, mainly because important terms remain highly ambiguous.

Cartographic efficiency

Much of the discussion about cartographic language is concerned with efficiency, i.e. the ability of the cartographer to devise the map so that it is easily and effectively used. The point is made by Koláčný, Ratajsky and Morrison. It is generally concluded that the cartographer's knowledge of visual perception and the problems of map use are inadequate, and that map communication would be improved by a more rigorous investigation of these aspects. This notion of 'efficiency' of course largely rests on the assumption that the communicated message can determine the response of the map user.

It is an interesting reflection on some academic approaches to cartography that the discussion of any knowledge of visual perception on the part of cartographers appears to be based on what is contained in published texts. Morrison (1976: 94) raises the question: 'What does the cartographer know about detection thresholds, discrimination thresholds, recognition thresholds and finally estimation thresholds?' and then follows this by stating that 'Cartographers have only begun to explore these areas.' Map users require maps, not theories, and cartographers spend their time making them. What they 'know' is incorporated in what they do, and the evidence is abundant. Maps are produced by taking a large number of specific decisions about content, form and design. The quality of these decisions as a whole reflects the cartographer's skill and knowledge. So far as I am aware, the Swiss cartographer Imhof did not publish any theories about visual perception, even though he produced many stimulating accounts of the principles and practice of map design. Are we to understand from this that his knowledge of detection, discrimination, etc., was inadequate? There are many maps in existence which demonstrate a very refined treatment of the problems of detection and discrimination. What the authors of some communication theories seem to be objecting to is that cartographic design has not been reduced to a set of formulae. There were no formulae for the design of the medieval cathedrals of Europe, but they remain momuments to human expression.

Writing in 1891, Andre observed that in regard to the area colours on a map, 'Tints that are of small extent may be a little exaggerated in intensity for the purpose of giving them greater distinctness.' In this remark he expressed what most cartographers know: that a given area symbol when applied to an area of small extent (a small visual target) has less contrast than when applied to a large area, and this can be compensated by slightly increasing the contrast with the background. It is doubtful if Andre had ever hear of psychophysics, but he knew about discrimination thresholds in practice. If the works of good cartographers are studied in detail, much can be learned about the problems of visual perception as applied to maps, and the correct use of the cartographic 'language'. People become fluent in a verbal language by using it, not simply by studying its grammar, and the acquisition of this fluency is strengthened if the best examples of the use of language are closely studied.

Such an argument is likely to be countered by the statement that no one knows how effective such maps really are because objective tests on their use have not been carried out. There is some truth in this, but there is also the problem of deciding whether a map made more effective for one user will necessarily satisfy another. In addition, there are large numbers of maps, in particular topographic maps, and nautical and aeronautical charts, which have been field tested by the users over long periods. It is difficult to believe that in such circumstances the map designs are seriously deficient. What is much more likely is that the maps could be improved from a user's point of view if the cartographic input was increased. This would inevitably make them more expensive, to which the users would object. After the first series of the Ordnance Survey 1 : 50 000 scale map was published, there was some criticism of the apparent similarity of the symbols for footpath and county boundary in some areas. Although a map user with good discrimination would not have been confused, the rapid response indicated that the users of such maps do not hesitate to complain if the map 'language' is not clear.

It is comparatively easy to find examples of poor or badly designed maps, which have not solved all the problems of discrimination, or in which the overall design is out of balance. But an explanation of why this should occur has to go beyond any question of communication theory, and take into account the many factors – especially those of time and cost – which circumscribe map production. The lack of formal rules for map design is not simply a consequence of cartographic incompetence, or a lack of interest in the map user. It simply reflects the sheer difficulty of deducing a set of rules, capable of universal application, from the specific solutions decided for particular maps. The complexity of obtaining information from a map is matched by the complexity of creating it. Cartography could benefit from specific advice which could be applied to make decisions about particular maps. But general recommendations about visual perception do not serve this purpose. Indeed, there is nothing whatever in the communication transmission theory that can be actually employed to improve a specific decision about the use of cartographic 'language'.

Cartography and map making

The problem of defining the relative positions and connections between map makers and map users raises some fundamental questions with regard to cartography, as well as communication. Map users are concerned with maps, not cartographic theory, and are affected by the information contained in them as well as its visual presentation. If a complete theory of communication is described as 'cartographic', does this mean that the term 'cartography' covers all aspects of the map, including its informational content? If the starting point in the communication system is the input of information, then presumably the quality, characteristics, availability and acquisition of that information are relevant to the successful functioning of the map.

In this respect the different interpretations of cartography are highly significant. If cartographers are essentially concerned with the preparation, design and construction of the graphic image of the map, then they deal directly only with part of the whole operation of map making. But if cartography extends across the whole of map making, so that it forms a complete 'system', then it also has to deal with the problems of data collection.

In the English-speaking world, the practice of cartography and the work of cartographers (how ever loosely the term may be used otherwise) led to the acceptance of the narrower definition of the subject as representing a coherent body of principles and practice. Most cartographers would not regard themselves as surveyors or photogrammetrists. They certainly do not regard themselves as geological or soil surveyors. The picture is further confused because many texts on cartography deal with the preparation of data from some secondary sources (especially statistical sources), but exclude the acquisition and preparation of data from primary sources (such as basic surveys). The 'thematic' maps frequently described in detail are almost entirely small-scale maps derived from statistics, although a geological map or a soil map is just as 'thematic'. Although this may simply reflect the interests of geographers, it suggests indirectly what is believed to be appropriate to the subject of cartography.

This confusion permeates the whole discussion on cartographic communication. On the one hand there are vague references to 'observing reality', as though this was an adequate account of the nature and problems of data collection. On the other hand, in the communication models the focus of attention is almost entirely on problems of visual perception. For cartographers to be concerned with visual perception is perfectly proper. But to assume that this is the central problem in the satisfaction of map users is another matter.

CHAPTER 9

Map-using operations

The terms 'map use' and 'map user' appear in a variety of contexts, and it is necessary to consider carefully their various applications. Although map users can be regarded collectively as the total range of people using maps, their interests and activities are diverse. Groups of map users may have interests in common, which may be identified by formal consultation with them, or anticipated by map makers. Some such groups may exert a strong influence on the creation of maps, especially if their consumption of map sheets is high, or if they are regarded as 'important' in the community. In some cases they may not be a cohesive body at all (e.g. motorists), even though they are numerically important as possible map purchasers. A special-purpose map is generally easier to deal with from the map maker's point of view, as the expected needs and interests of the prospective users, and possibly their level of skill, can be more easily summarized. Multi-purpose maps, including many special-subject maps, are open to a greater variety of potential uses, and frequently have to compromise between very different interests.

Although it is possible to regard map use in terms of groups of people with different interests, map use itself is essentially an individual activity. It can be collective, in the sense of groups of soldiers using the same map to reach an objective, but even then the process of extracting information from the map is individual. So far as map communication is concerned, the determination of scale and content has to be based on the needs of groups of map users, which is one factor in the 'input'. This aspect has to be distinguished in the communication process from the activity of individual map users, which is the operative part of the whole communication act.

A map-using operation, in the sense of one person's activity with a map, does not arise simply as the direct consequence of a map-making act. In the communication diagram, the arrows linking map maker, map and map user suggest a continuity of activity and a uniformity of intention which is quite at variance with what normally happens. Map use begins when a person becomes conscious of some particular problem which requires information for its solution, and realizes that this information could best be obtained from a map. This may be quite obvious to an experienced map user, but in many cases it does not happen automatically at all. There are thousands of motorists who find their way by following signposts and asking for directions, apparently unaware of, or indifferent to, the value of maps for such purposes.

The next stage is to formulate the enquiry in map terms, which must involve concepts of map type or subject, scale and area. This itself requires some understanding of what types of map either do exist or are likely to exist. A map user cannot operate on a nominal or theoretical map: she has to seek something concrete, whether or not it actually exists. Whether such a map exists, whether or not she can identify it, and whether or not it will be accessible to her, are all important factors in determining whether any 'communication' will take place. In this respect the map user is the initiator of the map-using operation: she does not simply 'receive' cartographic information.

Even if she reaches the point of obtaining the required map, she still has to devise her enquiry within the framework of the map information. Wanting to go somewhere does not in itself provide the clues as to how the map can be employed, or what combination of map information and other information will be necessary. The ability to 'read' the map also assumes that the map user understands what the map is being used for. Although a skilled map user may be able to derive some benefit from a poorly designed or out-of-date map (because attentive visual perception is highly flexible), the 'best' map will be of little value if the map user does not know how to consult it.

Map users and psychophysical testing

The same need to distinguish between map users as individuals and map users as collective bodies affects the application of any results from psychological tests of performance skills. Psychological experiments use groups of subjects or observers in order to accord with the scientific requirement for proper 'population sampling'. The end product of such a test is a statistical mean value. This may say something about the group as a whole, which is normally the objective of the test, but it may reveal little about the performance of any individual. He may be one of a group of map users, but he has to perform the map-using task by himself. The improvement of individual performance is more likely to take place through a better understanding of map structure and map characteristics than by regarding the user as a standardized performer of separate perceptual tasks.

Map content and cartographic elaboration

For many maps, the cartographic sophistication is a direct consequence of the map content. For example, many special-subject maps are devoted to complex distributions and many categories. These require an elaborate design specification if the information is to be expressed clearly and effectively. If this is constrained by extraneous circumstances (e.g. by limiting the number of printing colours) then it may prove to be impossible to present all the information sufficiently clearly, which in turn will pose problems in visual perception for the map user. Efficient communication, in this respect, requires not only theoretical cartographic knowledge, but depends on having the required technical facilities made available, in order to be able to make full use of the cartographic 'language'. In some cases there has to be a 'trade-off' between content and design, the content being reduced to the level where it can be presented effectively within the limits of a relatively simple product. The full use of the cartographic 'language' assumes a foundation of technical resources, as well as the availability of the required information. Providing the user with what is needed introduces technical and design factors which are a function of what the user can afford.

Map design and the map user

This particular relationship has many aspects and, as is quite correctly emphasized in many communication models, it needs critical attention. Several related questions emerge. To what extent can a cartographer anticipate the map user's skills and abilities? Is this a prime factor in cartographic design? If our present understanding of the map user's problems is insufficient, how can it be improved?

In dealing with the first problem, much depends on whether the map user is being directed towards some objective by a map author, or whether the map user initiates his own enquiry. If a map is an adjunct to a text, then it is reasonable for the author to assume that if the reader has the interest and the knowledge to comprehend the text, he should also be able to understand the concepts presented on the map. In this case the author has some idea of the intellectual level at which she is working, and can make some assumptions from that. This is not to say that problems of visual perception do not occur: only that a certain community of interest can be assumed between the author and the map user.

In those maps in which the user initiates his own enquiries, this is not the case. Once published, any map can be used by any person. Although it may be assumed that a geological map is primarily of interest to geologists, that is no restriction on its use by other people. A nautical chart may be used by a professional navigator, or by a recreational sailor with little knowledge of navigation. A topographic map may be used with great skill by a company commander or quite incompetently by a motorist with little knowledge of (or interest in) maps.

In general it is reasonable to say that the basic conditions which can make use possible should be satisfied by the cartographer. This is primarily a matter of detection and discrimination. There is no particular reason why this should not be so, as a cartographer operates with the same visual mechanism as other people. Discrimination itself requires a level of attention, and some understanding of what to look for, on the part of the map user. The person who does not notice that one line is thicker than another is not necessarily visually defective. He may not notice because he has not learned to notice, and is possibly unaware that a difference in line thickness is significant.

Problems in detection and discrimination are not limited to map users. More than one science teacher has lamented the inability of students to 'see what is there'. The matter is confused because it is difficult to establish whether an inability to perceive is due to lack of education or knowledge, or whether it may be due to different capacities of individuals. It is at least questionable whether a cartographer can be held responsible for either. As Petchenik (1977: 125) comments, 'If it begins to appear likely that each person is unique in the mental baggage he brings to the task of map reading, what is the poor cartographer to do with a map that is to be circulated among thousands of viewers?' Whether genetic or environmental factors are the main reasons for the different perceptual abilities of different individuals continues to be debated. As Gregory (1974: 51) points out, 'Traditionally the central problem in perception is how far the organisation is innate, and how far it is dependent on early learning.'

On this question the parallel with linguistic communication is interesting. If a reader is interested in studying some subject, or simply wants to 'find something out', he may select a book from a library shelf which appears to deal with the subject matter wanted. If it is too advanced for his level of understanding, he does not necessarily blame the author. If it is illegible, he may reasonably criticize the typographer and the printer. If a scientific author needs to use mathematics to express some statement properly, he assumes the capacity of the reader to understand, or his willingness to learn. The content and purpose of a map, and the level of 'language' needed to express it, go together. If a geological map with thirty or more rock categories is being produced, there is no way in which maximum contrast can be used for all the symbols. They are bound to require some sophisticated variations of pattern and colour differences, some of which may not be easily 'remembered'. This is a consequence, not of cartographic incompetence, but of the subject matter of the map.

The notion that map design can be reduced to satisfy the lowest level of perceptual competence would effectively rule out many maps and many map-using tasks. Although map design is carried out empirically, and is not based on a set of formal 'rules', the effective representation of the content, and the need to balance one element in the map against the others, means that the graphic variables of form, dimension and colour have to be fully employed. If a map is complex in its information structure, it is just as 'advanced' as a scientific text, and its use is bound to demand a high level of attention and perception on the part of the user.

Map design and cartographic objectives

If the cartographer cannot anticipate the individual needs and skills of map users, to what goal should map design be directed? Not only is this a critical question in the idea of a map communication system, but it also determines fundamentally what cartography (in the strict sense of the term) is about. Is it simply a matter of 'conveying information' or is it concerned with producing an orderly and systematic account which can be acted on by the user?

In order to deal with the question, it is necessary to restate some of the points made previously regarding perception and cognition.

Using maps is an aspect of human behaviour, and understanding how people behave is a basic aim of psychology. Discovering how people behave in using maps can be attempted in two ways. On the one hand, it can be approached by conducting psychological tests of the users' responses to individual map elements, which can be regarded as specific stimuli, derived from either 'real' maps, or other images resembling them. Therefore, tests dealing with detection or discrimination can be carried out, and the responses to different sets of stimuli observed. It is assumed that such psychological tests should lead to a better understanding of the mechanisms of visual perception. The results of such information could then be applied by the cartographer, leading to the production of maps that would be more satisfactory to the map user.

Much of this type of psychological investigation has been concerned with specifying detection thresholds (i.e. the minimum adequate stimulus under given conditions), or discrimination thresholds, (i.e. the minimum detectable difference). Such tests follow the usual scientific procedure to isolate a single variable. As a human observer responds to a stimulus (assuming that it is regarded as important) by using all available knowledge and clues in the visual scene, any group of observers will contain individual differences. Therefore they are highly unequal, and tests are devised to render previous knowledge or experience irrelevant, thus limiting the response to one variable at a time. The practice is described by Dobson (1977: 47) in stating that for a particular experiment 'the use of actual map patterns was inappropriate since the map reader could be influenced by prior knowledge of the spatial distribution of the phenomena.' The methodology is exactly described by Petchenik (1974): 'Research is conducted in a manner similar to that of the physical sciences, where one variable at a time is isolated and explored while all others are held constant.' This behaviouristic view is essentially a combination of 'mechanistic' and 'physiological' models of human communication.

On the other hand, tests can be conducted in which map users are invited to perform actual map-using tasks, and their behaviour and reactions observed. Such investigations are considerably more difficult to devise and analyse, and they can be illustrated by the studies of Hill and Burns (1978). Some investigators have even gone to the lengths of producing complete experimental maps and testing them, as reported by Taylor (1974). Hill and Burns (1978) explain that motivation and personal preferences and prejudices all affect map-reading performance. Sandford (1980, 1985) makes many interesting points about the use of school atlases by children, and P. Bailey (1984: 63) recounts his own experiences to emphasize that 'The most important observation is that I only began to learn how to use maps when I found my own reasons for doing so.' In general these investigations can be characterized as cognitive studies, in the sense that they are concerned with explaining or evaluating the performance of the whole person in real operations. The need for such studies is expressed by Neisser (1976: 7) in his statement that 'cognitive psychologists must make a greater effort to understand cognition as it occurs in the ordinary environment and in the context of natural purposeful activity.'

The notion that cartographic communication could be improved by learning from psychological tests of the first type rests on an assumption common to behaviourist theory. The view is clearly represented by Moles (1968: 3) in making a general statement of the aims of theoretical psychology: 'A theoretical psychology ought to develop along with experimental psychology and aim, on the basis of a normalised model of the human organism furnished by statistically cumulated experiments, to determine the individual's behavioural mechanisms, expressible in mathematical terms.' It is easy to see the parallel between the 'normalised model of the human organism' and the behaviour of the map user in the communication theory model. The same line of thought is expressed by Morrison (1978: 102) in his opinion that 'If functions defining map reading processes can be specified, certainly this information will be valuable in map making.'

Cognition and perception

In a long and sustained examination of the behaviourist approach, and the use of 'scientific' analogies, Oatley (1978) raises two main criticisms. In the first place, a description of the separate steps in visual information processing does not explain how it takes place, or what factors operate in controlling the mental processes, especially when isolated responses are examined under artificial conditions. In the second place, the criteria on which such tests are based are related primarily to the needs of psychological experiments, rather than to an overall understanding of visual perception. Oatley points out that the aims of scientific testing and scientific theory are based on the principles of repeatability, prediction and control. Repeatability is required to verify a scientific proposition. As Board says (1978: 41), 'The hallmark of a good theory is that it has stood up to empirical testing under replicable conditions.' If a theory is demonstrated by a given experiment, for example by testing the responses of a human observer to a stimulus, then it is proved by repeating it with other groups. If this is carried out conclusively, then it should be possible to predict the response to the particular stimuli. This therefore leads (at least in theory) to the ability to induce certain responses under set conditions.

This conception of the scientific experiment also fits in neatly with the theory of cartographic communication, in which an initiator (cartographer) endeavours to induce the desired responses in a 'receiver'. In this way the operation can be regarded as 'scientific', and in such a manner cartographic design can be converted from its present empirical basis to become a proper 'science'. This approach as a whole is challenged by Neisser (1976: 177), who concludes that 'The facts of cognition imply that the psychological manipulation of behaviour is bound to fail; it cannot lead to systematically predictable outcomes under ordinary cultural conditions.'

Oatley (1978: 230) comments that such experiments 'contain a strong commitment to the view that there are isolatable pieces of behaviour (responses) and isolatable causes of that behaviour (stimuli).' In arguing that quite different theories or analogies are needed to understand how the brain works (and therefore behaviour), he suggests that the accepted scientific approach is only valid in relation to certain phenomena, and that the analogy of perception with some sort of linear information processing is not necessarily fruitful. On this general point he maintains that:

> Scientific theories then apply to those aspects of the universe which are isolatable, stationary and repetitive. In those circumstances, by processes of conjecture and refutation, we can make successive approximations to, though not attain, descriptions of the universe which within their own terms are true. This, of course, puts psychology, where phenomena are not typically isolatable, stationary or repetitive, in a difficult position.

Neisser (1976: 8) makes the rather caustic comment that 'A satisfactory theory of human cognition can hardly be established by experiments that provide inexperienced subjects with brief opportunities to perform novel and meaningless tasks.' It is the capacity of the human perceiver to learn, and even to learn and change strategies during a task, which makes the deduction of any set of 'rules' so difficult.

A general dissatisfaction with the application of psychological testing to cartographic tasks was clearly expressed by Petchenik (1977: 117). In one of the most penetrating studies of the problems of visual perception and cartography, she identifies the central issue clearly:

> A considerable amount of perceptual research within the general framework of behavioural psychology has been conducted by cartographers during the last ten or fifteen years. However, as one reviews the findings of this research in connection with the problems encountered during the formal process of making maps, it doesn't seem to add up to much. No whole theory or set of principles, greater than the sum of the component parts, has emerged. Similarly, analytical attempts to deal with the notion of map reading have not led to any theoretical structures from which principles that would assist in the details of map design can be deduced. Clearly, map reading is more than just the cumulation of a number of simple perceptual comparisons of symbol size or value.

Oatley (1978) raises the same fundamental question in broader terms. He argues that in perceiving the real world, neither a pattern of excitation on the retina, nor a set of signals about edges and orientations from the ganglion cells, can describe what people 'see', which must involve some kind of internal mental representation. Petchenik (1977: 120) refers to essentially the same question in stating that 'Meaning seems to come from an all-at-once grasp of the relation of the stimulus to the reader's previous knowledge structures, rather than from a bit-by-bit build up.' The most difficult problem lies in attempting to define in what form an 'internal representation' exists. If this is gradually constructed on the basis of perception and experience, then presumably it must be particular to the individual, even though its limitations may apply to the whole species. It is in this sense that Oatley (1978: 229) maintains that 'Perception can be characterised as making meaningful sense of data collected from the outside world' and that perception itself involves 'the effort after meaning by the projection of aspects of our cognitive schemata on to the data' (Oatley, 1978: 225). The same general principle is emphasized by Young (1978: 123) in stating that

> We cannot possibly have enough storage space to build all possible programs for action of use each time we take in information. Instead we use overall hypotheses about what the world is like, and look for items that assist in carrying out whatever course of action we are engaged in. So perception is an active search for meaningful clues, and the brain builds programs to guide that search.

In this view all perception involves interpretation, and by definition emphasizes the cognitive elements rather than accepting processes such as detection, discrimination and identification as self-standing. In this respect, perception cannot be regarded as independent of the knowledge and mental state of the perceiver, and this point is also expressed by Salichtchev (1978: 98): 'The fourth stage – interpretation of the information obtained from the map – has as its chief aim the formation and expansion of ideas on mapped reality by enlisting the reader's prior experience and knowledge. But experience and knowledge vary from person to person.'

Although the application of the ideas of cognitive psychology to cartographic tasks was initially considered within the general framework of communication, the emphasis turned from treating the map and map use as a closed system to examining map interpretation as visual information processing. This stressed the importance of the knowledge and skill of the map user.

The basic concept was described by Olson in referring to an 'attempt to improve reader skills' (Olson, 1975: 94). Dobson (1979) extended this by producing a model of visual information processing applied to map interpretation, in which the physiological elements of the visual system are treated in relation to the cognitive processes. This was both interesting and unusual, for cognitive psychologists at that time tended to ignore the physiological nature of the brain, regarding the 'mind' as a separate entity. Perceptual and cognitive modelling was taken further by Eastman, who described information processing as 'one unified system of information capture and knowledge acquisition' (Eastman, 1985: 96).

Although this line of enquiry was a refreshing change from behaviouristic psychology, and made clear the complexity of map interpretation as visual information processing, the conclusions inevitably tended to result in generalities, such as the importance of map legibility, and the attitude and skill of the map user. Although enquiries into specific map-using tasks continue, the idea of constructing comprehensive cognitive models seems to have declined.

Given that a real difference of opinion exists between those who believe in the value of the application of psychological tests of the S–R type, and those who believe that perception cannot be examined on such a basis, a simple example from a fairly common design problem in cartography may help to illustrate the basic problem. On virtually all topographic maps, contour lines are included, frequently represented as continuous fine lines in a dark brown hue. Generally speaking, at least two types will be present: standard contours and index countours, the latter being slightly more prominent than the former. In terms of the intial stage of the perceptual response, the lines have to be detected and the difference between the two types discriminated. A normal gauge for a standard contour would be 0.15 mm, and for an index contour 0.25 mm. A psychological discussion would range over the question of the absolute threshold (the minimum thickness and colour of line detectable) and the differential threshold (the minimum difference in line thickness detectable). In practice, both lines are kept well above threshold, which of course is also affected by the particular hue of the lines and their contrast with the background. It is perfectly possible to distinguish between lines of 0.15 and 0.2 mm in thickness, but there is no need to make the task critical, and therefore the contrast is extended. A difference of 100 per cent would be too large, as it would make the index contours too pronounced. (This, of course, is a visual judgement.) In passing, it may be noted that the inclusion of index contours is purely to help the map user: there is no difference in the 'information'.

This sort of problem is familiar to many cartographers. Taken in isolation, for example in the legend, both lines could be made finer, and closer to absolute thresholds. On the map, some contour lines may be perceived against a white or uncoloured background thereby having maximum contrast. In other parts of the map there may be areas of woodland, represented in a continuous light green or a green pattern, which will reduce the contrast slightly. In still other areas there may be a combination of woodland and grey hill-shading, and perhaps many names in black. The contours may still be perceptible, but closer attention is needed to perceive them clearly. In a poorly designed map the contours may be so fine that they virtually disappear in some areas, although remaining visible in others. On the other hand, if they are made thicker in order to compensate for the decreased contrast in some areas, they may have the opposite effect and obtrude on other information.

This may be regarded as a perceptual problem, but a cognitive approach would have to take into account other factors. A map user simply looking for a place name or a road junction might be quite indifferent to the contours and pay no attention to them. Another map user trying to visualize the terrain might regard them as important, and wish that they were more prominent. How many factors would a 'scientific' test have to include in order to arrive at some 'measure' of the efficiency of the map design in relation to map use? There would have to be a different 'solution' for each combination of line thickness and hue, and each possible combination with other map symbols, and each change in the map user's requirements.

The importance of the cognitive approach can also be demonstrated from the same example. Being able to sense the brown lines, or distinguish between them, only leads to the conclusion, 'I can see two sorts of brown line.' This is not what is really meant by 'map reading'. Although the designation 'contour' will have been given in the legend, the term 'index contour' will not be explained, and it is assumed that the map user will learn to understand its function by inference from the map. If the map user cannot understand the meaning of 'contour', perception does not take place. Correct identification of 'contour' will enable the map user to determine elevation within certain limits. Yet 'reading a topographic map' is normally assumed to include the interpretation of slope, which is not directly given by contours at all, and is not part of the map content.

Even if a 'grammar' of cartographic design could be produced, it would not remove the condition that the map user must understand the symbol designations and learn to interpret their spatial locations correctly. This has to be assumed as part of map use. As Kolers (1973: 39) points out, 'a drawing of a circuit and a map of terrain are far more easily interpreted than are the corresponding words, for both represent spatial arrangements. When saying this, however, we assume that the interpreters are pictorially literate: that they know how to read pictures and maps.' Knowing how to read subsumes far more than being able to recognize the letters of the alphabet, or give the rules of grammar.

This discussion inevitably returns to the original question: what should be the objectives of cartographic design? As the cartographer cannot rearrange locational facts, but must endeavour to represent them correctly, then the context of a particular map has a profound effect on the perceptibility of the information contained in it. As the combination of content and representation can include an indefinite number of possibilities, its effects cannot be anticipated with certainty so far as the actions of any individual map user are concerned. The only way in which a 'perfect' map could be made would be to construct a map specifically for a single and thoroughly defined map-using task, on behalf of one particular map user. Even then, it would still depend on the user's own efforts.

What the cartographer should attempt is precisely what has been understood intuitively by generations of cartographers. Given some account of the purpose of the map, either by the map user or the map author, she seeks to represent the content by a systematic and consistent use of the graphic variables, adjusting these for the balance of the specific map, within the constraints imposed by the production circumstances. In cartographic circles, it is often observed that a good cartographer is one who can make a usable map within strict limitations of time and resources. Her attention is focused on the internal order and harmony of the map, in which the relative importance of different phenomena and classes are demonstrated by the graphic presentation. The difficulty of doing this in practice is well illustrated by Taylor's (1974) interesting account of the evaluation of an experimental map. The conclusion reached by Castner (l979a: 155) also confirms this view: 'About the only real design rule that emerges is to ensure the clarity and legibility of all map elements, for the most important processes that lead to specific task solutions are ultimately the cognitive ones.' If the cartographer succeeds in doing this – and frequently it is a difficult and complex task – then she will provide a basis on which map users can effectively work. The 'art' of cartography, so clearly referred to by Karssen (1980), is not simply an anachronism surviving from some pre-scientific era; it is an integral part of the cartographic process.

Map use and cartographic research

Research into different aspects of map-using operations and map users' problems is generally divided between two main types. The first can be described broadly as research into map user requirements, whereas the second is concerned primarily with investigating map-using tasks.

Both individual researchers and mapping organizations conduct enquiries into users' reactions to particular maps, or their proposals for modifying or improving them. Some national surveys, such as the Ordnance Survey, have arranged consultations with groups of users or have carried out research into user requirements. Many of these investigations involve surveys of users, usually on the basis of questionnaires. McGrath's (1971) work on national parks is a typical example. Others have tried to examine the content and design of particular types of maps, such as atlases (Hocking and Keller, 1992); the needs of particular groups of map users (Morrison, 1979); or the effectiveness of different versions of the same map information (Gill, 1993). For general topographic maps, which are the ones most thoroughly examined, the major difficulty is to relate the expectations of an individual user to the overall balance of the design and content of the map. Most users tend to see the map from the point of view of their particular informational requirements rather than as a whole. Inevitably, most map users have little comprehension of what is possible cartographically, and whether their criticism is due to the inadequacy of the map or their own lack of understanding. A degree of confusion about scale and generalization is typical. Where a specialized map is concerned then the map itself may be preceded by some attempt to identify the specific needs and interests of potential users, but it is usually difficult to obtain realistic suggestions from potential users without giving them a map to look at in the first place.

The second type of research has received more attention in recent years, although it is notoriously difficult to conduct. The main distinction is between research that concentrates on a particular map-using operation, and research that concentrates on a particular map-using group. The former is illustrated by Hill's (1974) studies of orthophotomaps, and the work of Hill and Burns (1978) on map-reading tasks in the field. For the latter, there have been a number of studies of children's performance in map reading, of which that by Carswell (1971: 44) is a good example. One conclusion reached by Carswell is illuminating: he suggests that 'as children move through the elementary grades their spatial abilities are suppressed if not extinguished.' Much of the evidence provided by such research indicates that most people are comparatively unaware of the nature and characteristics of maps, and that, in general, map-reading skills are low. In some cases at least, adequate instruction can rapidly make a significant difference. The inevitable conclusion is that any improvement in the use of maps must depend largely on increasing the general awareness of maps, and providing more and better instruction in map use. The lack of such awareness and instruction can largely be attributed to the general indifference to the graphic image and graphic symbols in normal educational systems. Literacy and numeracy are regarded as essential, but there are no incentives or awards for 'spatial ability'. Although cartographers can continue to argue in support of any attempts to change such attitudes, unfortunately it is not within the powers of cartographers to bring this about.

Although clearly there are no short-term solutions to this general problem, some suggestions could be made so far as instruction in map use is concerned. The evidence offered in the first few chapters indicates that effective map use depends largely on understanding the symbolic structure of the map as a means of representation, and the processes of visual perception involved in interpretation. Some attempt to explain in general how signs work, practice in visual discrimination, a simple description of the graphic variables, and an account of the effects of scale and generalization, might provide a more substantial basis for the development of map interpretation skills.

Given the speed with which many children have become accustomed to computer graphics, it is quite possible that a suitable package could demonstrate the relationship between map symbols and topographic features with which children are familiar, and then show how these are composed in a map. A series of images could demonstrate differences in scale, and the uses of maps at different scales. In this way the relationship between map structure and symbolization and the 'real' world could be presented, bridging the gap between what the child sees and how things are represented in maps. Even a small amount of instruction of this sort could provide a much sounder foundation for map users. Although the screen images would have different visual characteristics to printed maps, the fact that both forms exist would also help to prepare children for what will probably be a much wider use of digital and video map images in the future.

CHAPTER 10

The map as rhetoric

Concepts of rhetoric

Virtually all discussions of rhetoric deal with it in terms of language – initially the spoken language of the orator, and subsequently the written text. Whether it can be applied to other forms of symbolic representation and expression (such as the map) is by no means clear. Yet the power of the visual image in modern society would at least suggest that non-verbal communication should be influenced by the same factors and conditions as those of language.

In the most widespread and current usage of the term, rhetoric is commonly associated with the use of language to mislead, false ornamentation, or presentation without substance: hence the frequent use of the terms 'mere rhetoric' and 'empty rhetoric' as expressions of contemptuous dismissal. This is a consequence of the long period in which rhetoric declined into no more than 'eloquence' – the ability to deliver a speech or write a composition – depending solely on stylistic devices, and in the case of oratory, voice and gesture. This, of course, is a misinterpretation of the role of rhetoric as conceived in classical times, and even during the resurgent interest in rhetoric characteristic of both the eighteenth and nineteenth centuries in Europe and North America.

The range in significance between the two extremes – rhetoric as mere eloquence (which in fact is sophistry, not rhetoric), and rhetoric as the study of how all human communication operates – is so great that it is essential at the outset to attempt to clarify the different concepts of rhetoric, before attempting to discuss in what ways the practice and theory of rhetoric may be said to have cartographic implications.

It is also significant that a renewed interest in rhetoric should occur in the present period. In an age apparently preoccupied with both science and communication, it is perhaps inevitable that all forms of discourse should be considered in a new light. Despite the development of mass media and high-speed information transmission, in Britain at least a lack of skill in communication is advanced as a major criticism of the educational system. At the same time, dissatisfaction with the mechanistic model of human activity and response, with its attachment to a concept of objective 'scientific' communication, has led scholars in many fields to consider what affects communication in the broadest possible way. Philosophers, psychologists, linguists and historians have all been concerned with this re-evaluation. It is interesting that in the 1960s and 1970s, when geographers in particular were adopting scientific modelling as a basis for understanding communication, many other scholars in the field generally known as the human sciences were exploring a quite different approach.

In attempting to describe the potential value of the study of rhetoric, Vickers (1982: 248) raises a basic issue:

> The initial problem . . . concerns the nature of rhetoric, whether it is a discipline with a formal content proper to itself, not shared with other disciplines. The short answer to that question would be that it is not autotelic or self-sufficient, that it does not have a body of knowledge peculiar to itself . . . Since its field is human communication and human behaviour it links up with ethics, psychology, sociology, anthropology, politics, law. It is concerned with the methods of communication, expression, comprehension, persuasion, discussion – all the ways in which men and women interpret the experience of living.

In this sense, rhetoric is not a self-contained theory, but an examination of how communication between people takes place, taking into account not only the intentions of the writer or speaker, but also the reactions of the listener or reader.

Classical concepts of rhetoric

The present popular concept of rhetoric – and therefore the most common usage of the term – is very different to that conceived by Isocrates, examined by Aristotle, and later developed and amplified by Cicero and Quintilian in particular. According to the oft-quoted statement by Aristotle (see Roberts, 1984: 2153), 'the technical study of rhetoric is concerned with the modes of persuasion.' In this he emphasized that rhetoric is concerned not only with logical argument but also the emotive responses of the perceiver, and therefore separated rhetoric from both logic and dialectic. But he was careful to point out that 'its function is not simply to succeed in persuading, but rather to discover the persuasive facts in each case.' As Rosenfield (1971: 63) makes clear, Aristotle used the term both to describe rhetoric as an activity, a practice, and also to refer to it as a study of the principles by which successful communication can be achieved. Being aware that the art of successful communication could serve unworthy ends, Aristotle insisted that 'things that are true and things that are just have a natural tendency to prevail over their opposites' (Roberts, 1984: 2154). So although in some ways Aristotle took the subject matter of a discourse as self-evident, he did not divorce it from the intention of persuading the audience to appreciate the truth, or the need for a just decision.

The idea of rhetoric as an inquiry into how communication works, rather than as a technique of persuasion, was the one attacked by Plato, using the famous Socratic dialogues. The fact that this itself was deliberately misleading is argued by Vickers (1982). The distinction between rhetoric as conceived by Aristotle, and the superficial view which equates rhetoric with persuasion by any means, can be illustrated by the two English terms 'affect' and 'effect'. Any discourse is presumably intended to have an effect on its readers or audience. In the fullest sense, rhetoric is concerned with what *affects* this, taking into account that the secondary meaning of affect is to 'act upon, influence' (*Shorter Oxford English Dictionary*), or to 'touch the feelings of' (*Oxford Reference Dictionary*). In this sense, *'affect'* encompasses both the emotive as well as the factual circumstances. On the other hand, an *effect* is a 'result or consequence' (*Shorter Oxford English Dictionary*), and the corresponding verb is 'to bring about or accomplish'. Therefore *effectual* is what produces its intended effect, and to act *effectively* is to accomplish this decisively and completely. With his insistence on both the character of the orator, and the intelligence and understanding of the listeners, Aristotle was well aware that no discourse could be considered without taking into account how it would move the audience and arouse their feelings, that is, what would affect them. In the words of Cicero (Yonge, 1852: 527), 'he is the best orator who by speaking both teaches and delights, and moves the minds of his hearers.' For Aristotle, the orator must reason logically, understand human character, and appreciate the emotive power of words.

It is unfortunate that in English the term 'persuade' has prejudicial overtones. There is always at least the implication that whoever is being persuaded is being brought to some action or belief against their will or their best interests. In most respects, 'convince' is usually a more appropriate term.

Cicero and the parts of rhetoric

The five major divisions or parts described by Cicero and later amplified by Quintilian gave a much more formal and complete structure to the art of rhetoric than that advanced by Aristotle, and indeed these principles continued to be accepted as the basis for rhetoric for many centuries. In the *Treatise on Rhetorical Invention* (see Yonge, 1852), he listed them as *inventio, dispositio, elocutio, memoria and pronuntiatio*. Of these, *memoria* was inextricably linked with the need for an orator to memorize (and the listeners to remember) his speech, and diminished in significance as oratory itself became less important. The terms so used have to be interpreted in a very general manner. *Inventio* applied to the origination and formulation of the subject matter – an area given little attention by Aristotle. The arrangement of the discourse, *dispositio*, eventually led to extensive treatises on the various 'parts' which every address should contain, reaching its most absurd and artifical conventions with the elocutionists. According to Kennedy (1969: 77), Quintilian 'views arrangement as involving the whole question of the division of the subject and thus of the process of thought by which the orator brings his material into natural, coherent form.'

Elocutio, in distinction to its later interpretation, concerned the use of language to give form and expression to the subject matter, whereas *pronuntiatio* comprised the oral presentation, or what now might be called the delivery. It was the undue emphasis on presentation which in many ways led to the excessive concern with 'eloquence', and with questions of rhetorical manner, voice and gesture. To some extent this was partly a consequence of Quintilian's definition of rhetoric as 'the science of speaking well'. As McNally (1970: 71) explains, in the process of defining the four major definitions that have been applied to rhetoric, 'Subsequent approaches, unfortunately, tended to interpret "speaking well" in strictly aesthetic terms, so that rhetoric became advanced grammar – a study whereby grammatical correctness was advanced to stylistic excellence.' As Howell (1971) points out, for Cicero rhetoric accompanied grammar and logic as the essential bases of all linguistic communication of ideas.

Rhetoric and education

In medieval times, the basic *trivium* taught in the grammar schools included logic, grammar and rhetoric. These were the essential foundations of a classical education, and indeed they persisted in one form or another for several centuries. By the Middle Ages, the anonymous *Rhetorica ad C. Herennium* was generally employed as a sort of rule book, and is interesting not least because it was possibly the first attempt to provide a set of rules for an applied art. Although founded on Cicero in particular, it was concerned with practical instruction, not theory. For example, it stressed the objectives of correctness of language, clarity and appropriateness. Unfortunately, because of the detailed attention given to methods and 'figures', it was all too easy, as Dixon (1971) points out, to produce merely elaborate concoctions.

This separation of content from form left the way open for the possible treatment of rhetoric as merely style and ornament. Yet the accompanying idea, that rhetorical skill should be developed by a combination of analysis of texts on the one hand, and practice in composition on the other, eventually led to the modern concepts of practice in written composition and literary criticism. This form of study remains typical of the fine arts, whereby the student practices skills and also examines and sometimes even copies works of the masters.

The problem for the philosophers was to accommodate two apparently opposed means of persuasion. On the one hand, persuasion by force of argument leads to a concentration on logic and dialectic. On the other, an appeal to the senses raises questions of ethics. The outcome of the first is uncertain because readers or an audience may not accept or respond to rational argument. The second is limited because appeals to the senses can be exploited regardless of truth.

Given that a map combines factual information with symbolic graphic representation, it also raises a fundamental question about the nature of cartography. Although some maps are precisely informative and others vividly suggestive, it is possible to have both properties in the same map.

Although rhetoric was an important subject of study in the universities in the eighteenth century in both Britain and the United States, and many famous philosophers (such as Adam Smith) frequently lectured on it, it virtually disappeared during the nineteenth century, while a variety of schools of 'elocution' flourished. In Europe, some remnants of rhetoric continued to be taught until quite recent times, but most of these concentrated on 'figures' and formulae for composition, mostly to be learned by rote. The rise of the English 'plain style' movement, according to Steiner (1982: 23), 'was eventually to separate irrevocably poetic from non-poetic or scientific language (at least in the eyes of early modernists), since unambiguous correlation rather than similarity between word and thing was stressed.' Unambiguous correlation between symbol and 'message' was of course the objective of cartographic communication theory.

The institution and development of departments of English language and literature tended to take over the interest in linguistic communication and literary composition, so that eventually rhetoric ceased to have any significant function in education. In a sense, it survives in general education in the teaching of essay writing, or composition, in which the pupil is taught to generate ideas, arrange and develop them in proper order, and express them clearly.

The revival of rhetoric

The renewed interest in rhetoric, largely since the early 1960s, can only be accounted for by examining the major influences at work both in academic teaching and society at large. In the later nineteenth century the new field of psychology promised to satisfy what some had seen as the real objective of rhetoric. Through scientific study it should be possible to describe in objective terms how people reacted to external stimuli, and therefore how their behaviour could be predicted. This was in considerable contrast to Aristotle's insistence on the need for ethical values, for as Johnstone (1975) puts it, 'We see awesome persuasive powers in the hands of those who could not by any reckoning be counted as virtuous.' If the human mind could be examined 'scientifically', then it should be possible to ensure that an effect could be accomplished. Even so, the early stimulus and response theories of classical psychophysics, despite their interest, are no longer accepted as a basis for an explanation of human behaviour.

Although from the time of Bacon, a knowledge of 'truth' in the objective sense was thought to be achievable by logical argument and rational inquiry, the separation between 'art' and 'science' is a comparatively recent phenomenon. It is one which has particular significance in the study of cartography. It is noticeable that from Aristotle onwards, rhetoric was described as an 'art', whatever definition of it was assumed. The transition is revealed in the changing connotation of the term 'art'. Originally it served to distinguish those things created by human beings from those things which were part of the natural world, and for a long period, those things which were divinely ordained. All human products, mechanical as well as aesthetic, were regarded as the products of art. In this sense, 'artificial' was simply 'anything produced by human effort, not originating naturally' (*Oxford Reference Dictionary*), and an 'artifact' was any object made by human beings.

Yet the idea that art must be connected with skill is pervasive, so much so that the *Shorter Oxford English Dictionary* gives it as the primary definition, 'Skill as the result of knowledge or practice', and subsequently 'the application of skill to subjects of taste, as poetry, music, etc.' or even 'Skill applied to the arts of imitation and design.' This association can lead in turn to the notion that the chief objective of art is not utilitarian, but should be to provide sensuous stimulation or aesthetic pleasure, or to be added as an ornament to more utilitarian things. The separation between that which pleases in comparison with that which informs parallels the reduction of rhetoric to mere contrivance, as adopted by the many schools of eloquence.

Whether or not there can be human concepts of a natural or physical world devoid of any subjective appraisal, and whether information about such a world can be conveyed entirely systematically and objectively, is of course a leading question for modern society. It underpins the triumph of 'scientific' as the most valuable of all criteria in any form of communication, from a learned paper to a soap advertisement.

On the one hand, scientific thought is a function of a world view, a framework for understanding that characterizes any given period, and is itself dependent on a number of frequently unspoken assumptions. On the other, all scientific facts, as well as thoughts or ideas, have to be communicated between people, and this process depends upon the use of language, or some other sign system. As Vickers (1982: 248) says, 'Human communication is seldom that ideal of neutral performance which some opponents of rhetoric espouse, whereby one person addresses another without any thought of persuading, or convincing or moving them, with no concern whether understanding or agreement takes place. The neutral transmission of information – perception without judgment or evaluation – is a myth,' and he continues, 'we now know enough about the functioning of all sign systems to be aware of the inevitable complexities inherent in the act of interpretation' (Vickers, 1982: 249). In this sense, scientific knowledge can only be communicated through a symbolic structure which can be interpreted by others. It is this, in particular, which involves both science and art, even in exposition. In the words of Vickers (1982: 264) 'it seems unhelpful to continue to rank the arts and sciences in some sort of hierarchy: each has its own role, and the totality of human knowledge requires co-operation rather than aggression.' The obvious need for co-operation between science and art in cartography reflects this view.

Rhetoric and cartography

The connections between rhetoric and cartography need to be examined under several headings. In the first place, as cartography is essentially concerned with the making of maps, then the characteristics of maps, as combinations of information and representation, have to be considered. The map, like any other form of symbolic representation, has to be composed and presented.

Given that rhetoric has been fundamentally concerned with the use of language, in either a written or spoken text, the relationship between language and other symbolic representations is crucial. In such discussions, the terms text, discourse, exposition, narrative and communication are all employed, reflecting subtle distinctions in the nature and objectives of these uses of language. Their relationship to maps needs examination, but for the moment 'discourse' – that is the non-literary text – will suffice.

Intention

In making a speech it is clear that an orator has an intention or purpose, basically to convince the listeners that the propositions advanced are correct. This may range from no more than a description of some event to an attempt to change the opinions of the listeners on a matter of philosophical or social concern. Although some motivation must be present, it is often complex, and more than one motive may be involved in the same act.

The ways in which maps may be initiated, and therefore the purposes they are intended to serve, have already been mentioned briefly, to make clear that they cover a number of different objectives. At the basic level a map is devised to satisfy a need which is either known or presumed to exist. Although sometimes initiated by a individual, it is more likely to result from some government requirement or from a commercial company responding to a known or potential market. Specific intentions include the following, although the list is not exhaustive: to provide locational reference material, for example to accompany a travel guide, in which case a simple map will locate and name all the places referred to; to provide supporting evidence, for example in an article discussing a proposed alteration to a road network; to assist some action, such as the provision of a coastal nautical chart to aid safe navigation; to describe the location and distribution of some phenomenon, such as geological strata; to provide a general topographic description which will enable users to carry out a wide range of interpretations for specific needs. In some cases the prospective usage will be known; in other cases many types of use will be possible. A critical difference, however, between an oratorical address and a map is that the orator can respond to the reactions of the listeners, and therefore has a much more direct control over what takes place. A speech can be altered at the last minute in response to the reactions of the audience. This is not possible with other forms of communication. With maps, there is normally a much greater independence on the part of the user, for the use of a map will usually result from personal motivation, and always involves the user's interpretation.

Content and arrangement

Inventio, the instructive or informative aspect, that is the material itself, is the first consideration, but in a spoken or written text this has to be arranged in some way. For a map, once the topic, area and scale have been decided, then the material content can be assembled.

It is at this initial point that the difference between the map and a written text becomes significant. The content in a text can be arranged and ordered at will. It can be presented in outline in an introduction, developed in detail, and then summarized or concluded. Complex or important subject matter can be given more space than minor matters. Significant themes or ideas can be repeated.

One of the difficulties of dealing with the subject matter of a map is that the space available is determined by the scale of treatment. Complex or congested areas have to be dealt with in the same terms as areas carrying little information. Even though selective omission and symbol exaggeration are pushed to their limits, there is a maximum level of information density controlled by the limits of visual perception.

By controlling the order in which the subject matter is presented to the reader, the writer or speaker has a powerful weapon in deciding what impression is created. Some things can be glossed over, while others are treated in detail. With a two-dimensional image in which the positioning of the subject matter is determined by geographical location, the cartographer has no such freedom. He cannot stop the map user looking at some areas, and he cannot simply blank out other areas. Indeed, having to provide complete information for the whole geographical area of the map is one of the fundamental map-making conditions. The advantage of having a fixed scale, which is essential to many types of map use, has its disadvantages in the selection and treatment of content.

In this respect, the term 'order' or 'arrangement' in relation to a map is more applicable to the visual order which is achieved by the graphic expression. By balancing the relative contrast of major and minor subject matter, the most important elements are given emphasis and the minor or supporting elements kept in the background.

Expression

Elocutio refers to the use of the sign system to represent, to express, to give form to the content. Here again there are major differences between natural language and graphic representation. Linguistic expression can make use of a whole range of devices, such as repetition, antithesis, antimetabole, hyperbole and metaphor, and indeed the appropriate use and introduction of these was an important part of instruction in rhetoric. Of these, only metaphor can be said to have a clear cartographic equivalent.

Giving form to content is of course the 'art' of cartography, because the information has to be represented in some way. Although practices may evolve into conventions, as in any art, there is never an absolute control over the giving of form, which is why unusual or even revolutionary ideas can be introduced. It is this giving of form which makes possible understanding and interpretation by the map user. Finding the appropriate form – and there are always possible alternatives – largely determines whether or not a map is successful. Because it is a matter of skill and judgement it is essentially rhetorical.

It can also involve the elusive property of style, which certainly exists in maps, although it may be as difficult to describe in words as style in literary composition.

Expression, in turn, overlaps into the nature of aesthetic properties. Can these be summoned, as additional 'ornament', in the manner of oratory without substance, or do they arise from the cartographer's grasp of the relation between information and representation, inspired by originality within a conventional framework?

Giving form to content is fundamental in the use of any sign system. Understanding the nature and use of map symbols is just as important to the map user as understanding the typical devices of natural language. It is also clear that graphic symbols, as used in the map as a whole, can give an immediacy of impression, and can affect the senses, in a way which is only possible if language is used poetically. The nature of this impression is difficult to describe, which is one reason why map design, like other forms of art, is difficult to deal with in words.

Rhetorical effects

In addition to describing the components of any address, it was also accepted in classical rhetoric that the most successful orations would embrace *docere*, *movere* and *delectare*, i.e. they would inform, move the emotions, and delight the senses of the listeners.

If a map is simply regarded as 'information' then neither *movere* nor *delectare* would seem to play any role. Indeed they would be inimical to achieving a predetermined effect. In many types of oratory the need to involve the feelings of the audience was regarded as critical. But with any graphic representation, it is impossible to avoid creating such impressions whether or not they are intentional, and whether or not they produce any conscious response on the part of the user. People 'like' some maps and 'dislike' others. Generally this is consciously related to whether or not the particular map use involved has been successful. There are types of map which are obviously intended to create a favourable impression (such as those in tourist guides) in addition to giving highly selective information. One of the problems of hypsometric colour systems is that the visual impression created by the scheme may be at odds with the intended meaning. Although the colours describing different elevation zones are intended to be denotational, it is usually impossible to prevent the map user from responding to them at least subconsciously by adding naturalistic associations.

There are relatively few cases in which emotive power is used deliberately to mislead the user. It can happen (and does happen) if the methods of representation are inappropriate to the subject matter, but this is usually due to cartographic incompetence rather than deliberate intention. But whereas language depends on the responses to the meanings of words, the emotive effects of which have to be interpreted cognitively, a graphic representation necessarily creates a direct impression. How much this impression is sought, and how much it is an unintentional consequence of a particular selection of symbols, may be difficult to determine.

Finally, there is the question of whether the map is intended to please, or to delight, and whether such an attribute is a proper or necessary part of the map maker's intention. On the one hand, aesthetic properties can be regarded as roughly equivalent to literary eloquence or style in oratory: enjoyable in their own way, but contributing nothing to the understanding. Certainly, the ornamentation of the debased form of rhetoric has its cartographic counterpart. It is one of the arguments of the 'scientific' approach that cartographers can be motivated by a desire to make the map 'look nice', or to want to include some 'irrelevant' artistic quality.

On the other hand, it can be argued that all human artifacts have the capacity to arouse the sense of pleasure, and the enjoyment of human skill. This end of the rhetorical spectrum involves the connection between the map and works of art, accepting of course that art has a wider connotation than that frequently given by expressions such as 'fine' art or 'pure' art. It raises the question of whether a map can be held to be beautiful, or whether this is extraneous to, or part of, its informative function. The importance of this is argued at length by Morris (1982), in one of the very few works on the significance of the artistic element in cartography. She maintains that 'There is a constant debate regarding the respective places of science and art in cartography. The discussion almost always questions the place of art in the science and art of cartography, rather than the less egocentric and more revealing perspective of the place of cartography in art.'

In stressing the importance of the unity of content and form, rhetoric allows for the creative skill of the cartographer in composition and representation, while at the same time making clear that the symbolic presentation may be interpreted and responded to differently by different people. As a model it therefore finds a place for both the making and the use of maps which is more attuned to what actually takes place than models based on the application of science alone.

In considering these three fundamental properties, it is clear that the term 'map' encompasses a wide range of objects, a point which is often obscured by writing about maps as though they were all of a single uniform type. Clearly there are many maps that are primarily devoted to *docere*, and in which the emotive or aesthetic properties are of little or no significance. On the other hand, there are other maps which arouse expectations, which somehow create a symbolic representation that is both informative and vivid, and which are a delight both to behold and to use. At the farthest remove, some maps exist which are overly concerned with aesthetic pleasure, with elaborate arrangement and presentation, in which the subject matter may be inadequate, trivial or even misleading.

If rhetoric can be used as a means of understanding more about map making and the use of maps, then the means by which knowledge about place is presented in a specific two-dimensional form must be studied. In this respect the map has connections with, but differs from, both pictures and diagrams. Maps require both accepted conventions and systematic methods of representation. It raises the question of whether such conventions are universally apparent, or limited to those with a proper understanding. Can such representations be misunderstood or unsuccessful? If so, what is the cause of failure? Can the map deliver 'fact' separated from opinion or interpretation? Does the map succeed because it presents scientifically based knowledge, or because its visual order is convincing? To what extent is a successful map dependent on the feelings and individual reactions of the map user? The value of the study of rhetoric is precisely that it demands responses to such questions, however difficult it may be to provide answers.

CHAPTER 11

Rhetoric and communication

In many ways, the introduction of the idea of communication effectiveness into discussions of cartography, and the search for an explanatory model that would lead to an understanding of how this should be achieved, is itself a reintroduction of rhetoric under another name. Some writers would argue that the study of communication is simply a development of rhetoric in modern form. Others suggest that the idea of communication has replaced that of rhetoric. In pointing out that the problem of achieving effective communication has been discussed for over two millenia, Fisher (1978) uses the term communication to embrace rhetoric. Yet even in classical times, the notion that rhetorical skill could be acquired by learning a set of rules to cope with every kind of situation proved self-defeating. It is exactly what Cicero and Quintilian argued against. If the emphasis is given primarily to technical skill, then rhetoric is replaced by sophistry, by which the bad or the untrue can be made to appear as good or correct. This, of course, is exactly the false rhetoric of a great deal of modern advertising and political image making, and reflects the most common usage of the term. For if the means of achieving effective communication can be reduced to a formula, then they can be used to produce predictable effects.

In rhetorical terms, this attitude stresses *elocutio* as something which can be manipulated independently of *inventio* and *dispositio*. Cartographically, the equivalent of this emphasis on the choice of words, style and appearance is also sophistry, the means by which ends can be successfully achieved. This tendency is quite evident in the theorizing about cartographic communication, for although information and map are constantly referred to, that is the content and the structure, the real point of concentration is the design of the map symbols as the instrument by which the 'messages' are successfully conveyed. The linear nature of the typical communication diagram, with its sequence of separate processes, encourages this view of symbolic expression as something which can be considered separately from content and structure.

Rhetoric as skilled performance

If communication is regarded as the modern equivalent of rhetoric, then some of the considerations well understood in rhetoric need to be re-examined. Neither Cicero nor Quintilian could find perfect examples of oratory, nor could they find any single orator whose speeches were always at the highest level of excellence. They could find examples of good performances on particular occasions. In this they accepted that human performance, even by the most expert, was variable. They also accepted that relatively few people would have both the skill and the talent to become great orators. They recognized that oratory arose from different needs, and was performed under different circumstances, towards different ends. Whilst accepting the importance of an under-standing of the possible responses of the audience, they were aware that the good orator could not only make use of the predispositions of the audience, but could even change their minds.

Like the teaching of writing, or art, or acting, the teaching of rhetoric drew heavily on precept and examples. Initially these might be used as 'models' to be followed, but only as a means of developing individual ability. This stresses theoretical learning or study in combination with personal performance and practice. Writing, speaking, acting, painting – none of these can be acquired simply as a set of rule-governed techniques, even though systematic and orderly study is necessary. Without the effort of participation the skill is impossible to learn. The ultimate objective is to be able to combine the intention, the content, the structure and the expression into a coherent unit. Parkhurst (1898) makes the point in reference to composition in painting:

> Naturally, in dealing with a thing like this, which is the very essence of art, rules are of very little use. Ability in composition may be acquired when it is not natural, but it calls for a continuous training of the sense of proportion and arrangement, just as the development of any other ability calls for training.
>
> The best thing you can do is to study good examples, and try to appreciate, not only their beauty, but how and why they are beautiful.

When Koláčný (1971) drew attention to the importance of correct, legible and up-to-date maps in assisting geographical understanding, and the need to produce maps properly related to the requirements of map users, his comments were more an indictment of the nature of education and society than of cartography itself. This impetus could have been followed in many ways. It so happened that at that time it was interpreted as leading into communication modelling, but in different circumstances or at a different time it might have been pursued along other lines. As Hübner (1983: 214) says, 'the idea of scientific progress has been universally extended to the point where it is thought to encompass almost everything.' The way chosen reflected the overwhelming influence of the scientific method at that time. The constant reference to knowledge, and the absence of any mention of skill, was characteristic of this approach. Yet the 'art' of rhetoric, less pretentious in its scope, offers a quite different level of understanding.

Content and form

The idea that a map is a means of communication is a perfectly good starting point. The problem of being content with such undefined generalities as 'map', 'information', 'cartographer', 'reality', etc., is that the huge diversity of map making, maps and map use is ignored. In so doing, it fails to reveal the real nature of cartographic work and cartographic problems.

The study of rhetoric makes clear that human communication ranges over a wide spectrum, from the emotive/aesthetic intention of fine art to the rational investigations of science. An applied art and science, such as cartography, also ranges between these extremes. Whereas the philosophers generally sought to avoid metaphor, the nature of language makes this impossible, a fact increasingly recognized by scientists themselves. Maps are not uniform objects, even though they have a specific structure as a means of representation. The large-scale urban plan, depending as it does on highly abstract delineation, lies at one extreme. It is the nearest the map comes to being a systematic and unemotional instrument, devoid of any attempt at elegance, and so satisfying a minimal practical need. At the same time it is a poor representation of the 'reality' of the built-up area, as it exists in four dimensions. In order to be exact in two dimensions, it sacrifices everything else. Highly indexical in character, it directs attention to significant 'lines' at ground level. Given its conventions, and the particular methods of surveying, it would be possible for two such plans, at the same scale and of the same area, to be produced separately, and to be almost identical.

Compared with this, the small-scale map of a continent is a summary interpretation, founded on a judgement of what is held to be important, offering an infinite number of possible variations in interpretation and presentation. It cannot even be understood without imagination, for it presents in concrete terms something that is not directly perceivable. To what extent it is understood is a function of a whole complex of knowledge and attitude on the part of the map user. Because of the inevitable distortions inherent in its structure, its correspondence with the real world is highly artificial. If this is not appreciated, it can be very misleading. Misunderstanding is part and parcel of all human communication; so much so that Richards (1936) regarded it as unavoidable.

The fundamental characteristic of rhetoric is the insistence on the unity of content and form. This is what all the arts demonstrate. A map is a visually ordered structure, not a collection of messages separately conceived. Every real map poses a different problem. It may be approached on the basis of known conventions and practices which are familiar to map maker and user alike, but in all cases the informational content and the graphic representation have to be managed together.

Sless (1986) points out that despite the prolonged effort to teach the basic skills of writing, reading and comprehension in modern society, there is considerable evidence that about one-quarter of Britain's adults do not progress beyond a reading age of thirteen, and that in the United States more than 90 per cent of television viewers misunderstand a quarter to a third of what they see and hear. Even with news broadcasts, it seems that factual presentations are widely misunderstood. Given that understanding graphic means of communication receives very little consideration in our educational systems, it is naive to believe that changing methods of representation, or adopting new 'techniques', will substantially improve the understanding of the generality of map users.

Rhetoric and the art of cartography

By studying rhetoric, a number of significant elements in cartography can be seen not as aberrations from some scientific system, but as proper to any form of communication. Maps, like other artifacts, cover a great range of objects, differing in their complexity and the purposes they serve. Cartography can operate with simple terms, or elaborate expressions, and must encompass both. Content, structure and form, (i.e. *inventio*, *dispositio* and *elocutio*), have to be considered as a whole, for the harmonization of these is essential in any art. This requires skill based on practice, and at the highest level depends on individual talent. Although general principles can be studied, and conventions are a necessary part of the structure of maps, every artistic work is an individual creation, and therefore a particular case.

The history of rhetoric shows that the concept of what is regarded as proper to communication varies with the social conditions of the time. Historically it has ranged from an emphasis on plain speaking to a fascination with eloquence for its own sake. Whatever the predominant concern of the times, both elements – objective deliberation and eloquent persuasion – are present. It is the tension between them, the need both to inform and to express, that gives the possibility of outstanding work on those infrequent occasions when both these objectives are satisfied simultaneously. It is the possibility of achieving this, even with mundane things, that makes art so important.

How ever it was formulated at a particular time, rhetoric has always been described as an art, and like all arts it depends on skill. Competence in the production of a text is assumed to be within the possibility of every educated person. Competence in representational drawing could also be achieved if it received a fraction of the attention given to the study of language. Linguistic competence is regarded as a basic objective in education. Excellence is inevitably restricted to a few. This indeed separates the everyday use of language in written texts from the achievements of literature, even though the subject matter may be similar. Many people never master the problems of written composition, even though they can admire it when it is well done. Others are simply indifferent. This applies at all levels of art, whether in composition or performance. There are skilled performers who become engrossed in the sophisticated manipulation of the medium; there are others who develop a particular style, and remain content with that. There are still others who constantly strive to achieve something that will fully satisfy their intentions, without ever quite realizing it in full. Practising cartographers are themselves aware of these feelings of satisfaction, delight and despair. Most cartographers, like painters, are always conscious of the level of their own performance, and react to it.

As with writing or drawing, cartography can be practised at an elementary level, and in many cases this is accepted as sufficient. The world of maps includes a large proportion of usable products, much larger than the number of maps that can be regarded as excellent. Many cartographers, including those who make maps without being aware of cartography as a specialized field, work within the general framework of competence, either from choice or because of external limitations. Given such considerations, the parallel between cartography and graphic art can be explored to illuminate both the similarities and the differences. This is not to suggest that painting as a fine art provides a model for cartographic practice; only that it helps to reveal factors which are outside the scope of scientific methodology.

The other important characteristic of rhetoric is its openness. It encourages any exploration of theory or practice which may help in improving skill. In this respect there is no reason why interested individuals should not investigate psychophysical testing, cognition, or any other field, if they feel that this may improve understanding. At the same time, there are other fields of human experience, especially in the arts, which can be fruitfully examined. Research in one does not preclude the exploration of the other. A concern with science is proper. An insistence that *only* scientific methods are of value refuses any place to the other, equally important field of knowledge and experience.

Structure and expression

In any work of art, these two aspects have to be dealt with together. Even though in a representational painting the point of view and the overall content may be pre-considered, as the expression of details is realized it is done within a framework of overall intention. As such details are encountered, they may themselves lead to modifications of the composition. With a map, once it has been conceived, the actual positions of the elements being represented are fixed. In this respect, cartography is circumscribed, not free. Although this fixed element in the composition introduces a different set of problems to those encountered in representational painting, in both cases it is not within the powers of the artist to simply rearrange or displace the subject matter. The painter may choose the landscape scene, omit insignificant or distracting features, and concentrate on a personal interpretation. But the objects still have to be recognizable and correctly placed.

The other similarity lies in the need to balance one element against another in the creation of visual order. Cartographically, the choice of individual symbols in terms of their graphic specification has to be related to the overall composition and the effect that is being sought. In many cases this is not carried out with any sense of it being a special task; few cartographers do it consciously, and they have not necessarily reflected on it as a particular problem. Neither do painters necessarily think about techniques while they are painting.

Individual style is affected not only by the subject matter of the map but by the way in which the various elements are both contrasted and harmonized. Just as composition is regarded as the most difficult part of landscape painting, so it provides the most difficult problems in cartographic representation. This skill in imagination is the most distinctive feature of the good cartographer. As the appearance of the finished map evolves gradually during the production of the work, it is normally the case that the original intention has to be modified at the first colour proof stage, or the first complete visual display, for it is at this point that the graphic expression becomes visible. Only with very simple maps is it likely that the original composition and design can remain unaltered. At the same time it demands a level of self-scrutiny by the cartographer – an honest appraisal of success or failure – which lies close to the constant self-scrutiny of the artist.

Given these difficulties, it is not surprising that despite the emphasis in communication theory on the satisfaction of the user and 'effective communication', the practising cartographer is essentially preoccupied with giving form to content, i.e. with the problem of representation itself. The act of creating and producing focuses attention and demands concentration on the internal problems of visual order. As Eastman (1985: 100) says, a good map 'facilitates the reader's organisational exploration.' Unless this order is achieved, the possibility of communication does not arise. Nor can it be accepted that what one user may regard as a 'good' map will necessarily arouse the same response in another. The writer may well be aware of her readership, and indeed may be aiming at a certain market: but what she has to generate is the expression of the content. It is in this respect that working cartographers of my acquaintance find communication theory so remote: the generalities of 'message' and 'transmission' seem to have no connection with what cartographers are engaged in doing.

In studying cartography, the beginner has to start with discrete problems, dealt with individually. Gradually the ability to compose the map as a whole is developed. Eventually he is able to look at the content, in whatever form it is available, and realize immediately the central problems in representation and some possible solutions. Along with this may come the realization that a specific design is required, not imitating in any exact way any previous experience or model. This capacity cannot be analysed or defined on the basis of the rules which the beginner learned, for these are no longer involved. Nor does the expert gradually work through these rules in order to arrive at the decision for action. The reaction to the problem is that the irrelevant elements – those which can be dealt with without difficulty – are dismissed, and the critical point ascertained.

Art lies where skill is allied to creativity. The connections between the content, the form and the means of production are of course directly evident to the artist. The painter is actively engaged with the medium at a personal level, not a theoretical one. Certainly the painter will make use of a theoretical knowledge of colour, for example, but only as a basis on which to develop skill. The two are intimately connected. Just as the classical orator studied excellent performances as models, so the painter studies artistic works, practices particular skills, experiments with different forms. It is a mental and psychological world far removed from the arid and impersonal notions of 'messages' and their transmission. Like the orator, the painter can never be sure that what he produces will actually achieve his intentions: there is the elation of success and the despondency of failure.

In all the arts, the ability to execute must be combined with the imagination to conceive. They are not distinct and successive phases. In rhetoric, practice and study can be personally developed to the point where they provide an immediate understanding of what is needed and how to express it. At this level it is individual and inimitable. The art of cartography is not something which can be reduced to a set of rules or a succession of repetitions. Like all other arts, it requires a developed skill and a creative imagination, working within the limits of a given structure and the possibilities of a particular medium. As such it is not a minor element which can be attached to a 'system' or added as decoration for persuasive effect. It is fundamental to cartography itself.

CHAPTER 12

The map as a text

The idea of using natural language as a model for the study of cartographic signs has already been mentioned, and indeed the idea is given a certain credence by the frequent references to map 'reading'. As the map presents an account of the distribution of certain phenomena over a particular area of the Earth's surface, it can be likened to a descriptive written text. Although the fundamental distinction between the linear arrangement of language and the two-dimensional display of any graphic image is quite evident, the presumed equivalence between map and written text is a common characteristic of much theorizing about cartographic communication. The linear sequences of most of the cartographic communication diagrams, whether intentional or not, demonstrate this presumption.

The use of language as the model for the analysis of structure and meaning has been repeatedly challenged in relation to many fields of activity. Wollheim (1987: 44), for example, is opposed to 'all those schools of contemporary thinking which propose to explain pictorial meaning in terms like rule, convention, symbol system, or which in effect assimilate pictorial meaning to something very different, which is linguistic meaning.' Similarly, although in a quite different context, Tyler (1986: 40) maintains that 'Writing is the means for a systematic separation of form and content more pervasive than that conditioned by pictorial representation, and more accessible than anything provided by hearing or touch. Writing expresses the separation of form and content visually, and promotes our consciousness of it.' Casey (1971–72: 197) questions 'whether other forms of human expression depend to such a degree as does language on determinate rules, contents or contexts,' and goes on to suggest that 'The felt surface of language does not in itself typically engage our attention; what is presented to us phonically or in print – the "vehicle" – is normally surpassed and even suppressed as we attend to the specific meaning conveyed by this vehicle.' This separation of form and content, the fact that the visual response to language is concerned only with meaning, not affected by the appearance or colour or pattern of the type, emphasizes a critical difference between looking at a map and reading a text. The same point is made in a different way by Eco (1985: 178): 'verbal language is perhaps the only semiotic system in which one gets the impression that to a single expression corresponds a single content unit.'

The difficulty engendered by a preoccupation with language and learning from written texts is quite evident in the teaching of cartography. Students presented with a map to examine immediately focus on the search for meaning expressible in linguistic terms. The great problem is to make them sufficiently conscious of the connection between form and content, or even to make them react to what is visually before them. As Kittay (1987: 5) says, it is necessary to understand that 'we think *in language* just as the artist expresses herself *in paint*; that we understand that language is not merely a conduit. . . . But it is the view of language as conduit that has prevailed.'

The map as a text

Whatever viewpoint is adopted regarding any of the theories of semiotics, or the degree to which they provide a satisfactory basis for explaining human communication, it is evident that all such linguistic analyses have one element in common: they seek to unravel the components of any system by analysis, by looking for the units and the rules which govern their combination. It is in this area that the use of linguistic structure as an analogy for other forms of communication is most open to criticism. This type of analysis often appears to be deficient because although it may discover the constituents of any discourse, it fails to reveal how it is actually put together. In this respect Gray (1977: 213) argues that

> rhetoric is considerably more generative, more helpful in the actual inventing or creating or composing of discourse, than any generative grammar has proven itself to be. There are many different systems of linguistic analysis, of breaking down into constituent elements already existing compositions. But it does not follow, and it has not followed, that these are necessarily a help in creating compositions from scratch.

This dilemma is also apparent in all critical studies of the arts. Although they seek to describe in words the properties by which works of art are distinguished, and to compare the merits of one work with another, it is extremely difficult to express these matters verbally. What characterizes artistic works is of a different order to what can be fully dealt with in words. It is a point of some importance in regard to map making, in so far as some theories of cartographic communication purported to establish not only how maps operate in conveying information, but were intended to be used as a basis for research into rules for map design.

In several recent studies the importance of examining the structure of a text as a whole has been stressed. If for the moment it can be accepted that a map is a kind of text, a complete composition in itself which has content, form and intention, then the emphasis on this point leads back to the study of rhetoric. This contrasts with the general approach to the linguistic analogy characteristic of most theories of cartographic communication, which, at least in the past, have concentrated primarily on the dissection of language into its parts, rather than its composition into wholes. Rose (1986) maintains that the reductionist methodology insists on an analytical procedure, isolating each element and dealing with only one variable at a time. This could well be applied to the notion of cartographic communication when conceived as the isolation of fragmentary messages being transmitted from senders to receivers, as a description of what actually happens in the creation and use of a map. As Reid (1969: 235) points out, 'In science one can and often has to do one thing at a time. In art, the one is involved in the many and the many in the one, and this demanding unity dominates.' What the study of rhetoric emphasizes is that the whole structure, its composition, subject matter, means of expression, style and emotive effect on the intended audience have to be considered as a whole. In any rhetorical work, although common elements can be distinguished, there are no absolute rules for putting it together, a point constantly emphasized by both Cicero and Quintilian. In any analysis of the map, this rhetorical view encompasses the map as an artistic, as well as a scientific work.

Semiotics and art

The question of how works of art are composed and appreciated, and the relations between the artist and the perceiver, are of critical interest to art historians and critics, as well as literary scholars. The notion that meaning is conveyed by arranging a series of comprehensible individual signs in an appropriate order (like a code) is inadequate as a description of either literature or graphic art. It ignores what Murray (1986: 11) refers to as 'imaginative thinking; the rhetorical, aesthetic, creative, ordinary dimension of language.' The idea is scornfully dismissed by Richards (1936: 9) in saying that 'to study the efficiency of language and its conditions, we have to renounce, for a while, the view that words just have their meanings, and that what a discourse does is to be explained as a composition of those meanings – as a wall can be represented as a composition of its bricks.' And he continues 'any part of a discourse, in the last resort, does what it does only because the other parts of the surroundings, uttered or unuttered discourse and its conditions are what they are.' The comment makes a point which is central to cartography, because of course the 'spaces' in a map (the unuttered discourse) are just as important in the total meaning as the specific symbols.

According to Lanigan (1972), a rhetorical statement is a mixture of scientific and emotive relationships, and so signs are used in both a subjective as well as an objective sense. Although the ordered two-dimensional plane of a map is nominally objective, in the sense that its correspondence with the 'real' world can be rationally defined, it may contain other signs, the forms and colours of which are essentially metaphorical, and have no objective connection with the subject matter.

In the many theories which have been advanced in discussions of the philosophical basis of art, certain components or aspects constantly emerge. One of these is the notion of representation, the way in which any work of art imitates or describes its chosen subject matter.

Imitation and representation

The relationship between the external world and its representation is most evident in the pictorial arts, although 'setting the scene' and creating 'atmosphere' are significant in many literary and musical works as well. The way in which a work of art may appear to imitate appearances has been the subject of many investigations: the classic study by Gombrich, *Art and Illusion* (1977) traces them in detail. Gombrich (1977: 21) stresses the role of perception, both in the artist's interpretation and in the viewer's response, even though at present there is no complete understanding of how the processes of visual perception operate. In his words, 'We have to get down to analysing afresh, in psychological terms, what is actually involved in the process of image making and image reading.'

One fact which is immediately evident in any graphic representation is that it is impossible to separate the content and the form. The content is only available when expressed in a particular form; there is no abstract or conceptual equivalent to the work itself. Although in language we can refer to something called a 'map of the world', any specific map is only one possible interpretation based on one constructional convention of representing a three-dimensional curved surface in two dimensions, with one particular generalization and one set of conventional symbols.'

The distinction between discursive forms and presentational forms is a major element in Langer's several works on symbolism in art. According to Innis (1985: 87), in Langer's view, discursive forms 'are marked by linearity, individual and isolatable signifying units, a syntax, and the possibility of translation. Presentational forms, however, are meant to be perceived as a whole *Gestalten*.' He points out that Langer does not pursue the semiotic analysis of non-linguistic analogies of phonemes, words, sentences, etc., but regards a work of art as a single symbol. This fundamental distinction between presentational and discursive forms is also taken up by Robinson and Petchenik (1976), who argue that the graphic image does not have any ordered syntax which corresponds to that of language.

Not surprisingly, the difference between language and 'visual language' has been stressed by many writers on graphics and the pictorial arts. Bonin (1975: 19) explains that 'In visual language the process of comprehension is the opposite of that followed in verbal language. Verbal language proceeds from a construction having as its base the rules of grammar, which are defined and known.' He goes on that 'So, it is by habit, using the false analogy of verbal language, that the problems of the composition of maps and diagrams are posed.'

Perkins and Leondar (1977: 6) maintain that 'It is illuminating to think of symbol systems as languages provided that the limitations of that analogy are acknowledged.' Schlichtmann (1985: 23) prefaces his remarks on map symbolism with the caveat that 'Metaphorical terms like "map language" and "cartographic language" might suggest facile analogies between language and map symbolism and are therefore better avoided.' Woodfield (1986: 357) suggests that 'The notion that a science of signs based upon a linguistic paradigm could provide a viable basis for the analysis of the visual arts is becoming increasingly fashionable,' before going on to criticize it in detail.

Semiotics and maps

Despite the large number of papers dealing with communication in cartography, relatively few have pursued in detail the analysis of map symbols, and the relationships between map symbols and semiotic theory. Of course, map makers, like artists, do not need to study a set of rules, and can even acquire the necessary conventions by acquaintance, rather than theory. It also demonstrates that map makers are concerned with making things, not explaining how they do so. On the other hand, it is evident that many ordinary map users, given that they are prepared to attend to the use of a map seriously and are normally intelligent, do manage to obtain information from a great variety of maps, despite the lack of any formal knowledge of their structure. It is doubtful if the user of a road map, for example, consciously considers what the lines on the map represent. She may approve or disapprove of the map; she may find it easy or difficult; she may even like or dislike it; but she is unlikely to look beyond the satisfaction of immediate needs. Unlike the reading of a scientific or mathematical text, using a map does not necessarily require a large range of specific prior knowledge of the 'rules' and terms. On the other hand, the use of a specialized map does require the map user to have a degree of understanding of the subject matter or phenomena represented. Like a picture, the map can be looked at with varying degrees of understanding, comprehension or appreciation. It is not the case that all map users are put off by difficulties; for many purposes, reading the map is surprisingly easy.

The same point is made by Blocker (1979) in relation to pictures. He maintains that 'One important *difference* between artistic representation and a natural language is the ease with which we are able to 'read' drawings once we understand the basic conventions. Knowing the word *cat* doesn't help me to understand what *dog* means, but once I understand a line drawing of a cat, I can immediately, without any additional tutoring, understand a line drawing of a dog, elephant or horse.' And he continues, 'Language representation is almost entirely conventional, artistic representation only partly so. But because pictorial representation is partly conventional, there can never be a purely objective artistic representation of the world.'

Cartography and the linguistic analogy

Despite a few attempts to examine maps on the basis of the linguistic analogy, the relationship between a map and natural language poses some basic difficulties. Certainly a map can be looked at as a whole, but it can also be used as a source of bits of discrete information, even though the essential fact that these bits are placed in a larger whole is what makes a map specifically useful. Whilst accepting the limitations of the language analogy, Head (1984) explores the possibility of analysing the map on the basis of language, or linguistic semiology. He takes the terms and concepts advanced by linguistics (phoneme, morpheme, word, syntagm, etc.) and attempts to see if there are any equivalents in map symbolism. In this he makes the point that although the map, like a picture, may be perceived as a whole, it is also scrutinized (like a picture) by scanning parts of it; that is, by processing it over time. The objective of this scrutiny may be individual things, or the relations between things.

In a text, the words themselves are not 'looked for', even though they have to be read. With a map, particular symbols, representing specific information, may be sought within the larger context. Head (1984) draws attention to the distinction expressed by Petchenik (1979) between place and space; between the use of the map to define the location of an object, and the use of the map to understand the relationships in space between objects. What is clear is that it is the map's capacity to represent systematically an area of the Earth's surface and of phenomena related to it which gives meaning to even the most simple task of locating a feature. It is also clear that although some parallels may be found between linguistic units and map symbol units, the differences are greater than the similarities. The further consequence of this is that by starting with the linguistic analogy, points which are significant cartographically may be missed entirely.

Where these analyses get into difficulties is in examining the relationships between the map as a designed object, and the interpretation by the user. When Head (1984: 15) quotes Bonin in describing 'maps designed for this function as *cartes à lire*', he also implies a correspondence between the map type as something intentionally devised by the map maker, and the level of reading or interpretation to which it is suited. Yet it is clear that the vast majority of maps can be interpreted at different levels, depending on the needs and objectives of different users. Levels of interpretation may be assisted by symbol design and map composition, but it is not normally within the power of the cartographer to provide a 'text' which can only be interpreted in a certain way. If this really is the objective, then the map would have to be limited to a small number of specific items to demonstrate only one statement. The same is true of all literature and pictorial art, and indeed the problem of meaning and interpretation is equally important in art criticism.

Language and text

In considering the many accounts of the nature and use of language, one factor is immediately apparent: the variety of terms used to refer to different types of linguistic usage. This stands in sharp contrast to the limitations of the single term 'map', despite the fact that maps are just as varied in content, expression and objective as linguistic works. Given the fundamental distinction between prose and poetry, and accepting for the moment that the types of writing of immediate concern could be placed under prose, the use of the terms text, discourse, exposition, narrative, description and even communication suggest a great range of intentional and compositional differences. They all seem to have some bearing on any analysis of the full meaning of 'map'.

Exposition seems to be reserved for the most ordinary or matter-of-fact attempt at description or explanation, offering nothing of aesthetic interest. The nearest cartographic parallel is probably the line plan, which 'refers to' something that can be identified and located on the ground, but otherwise says little or nothing about it. It is the use of map 'language' at the most basic level.

White (1957) makes the distinction between discourse and narrative. He explains that whereas discourse is personal, declaring the presence of a human author, narrative is characterized by the omission of any reference to the narrator. In this sense the 'impersonal' account, offered as factual description, is certainly how most maps are presented. Yet nothing produced on the basis of human judgement and intention can possibly be brought into existence impersonally.

Discourse is a term widely used to refer to non-literary writing as distinct from literature. Yet it contains a suggestion of personal origination and intention. There is at least the implication of individual speech or writing, as either monologue or dialogue. Yet the meaning of discourse appears to be so loose that McNally (1970: 77) finds it necessary to examine it in some detail in relation to the use of signs and rhetoric. Taking Morris's three divisions of signs – semantics, syntactics and pragmatics – as the starting point, he explains that 'if it can be determined that the "sense" of a piece of discourse resides chiefly in factual reportage and inference, we may call the discourse "semantically oriented"; if in elaborating the implications of a set of formal or aesthetic "premises" it is syntactically oriented; and if its meaningfulness consists chiefly in the address to the interpreter's responses, it is "pragmatically oriented discourse".' McNally concludes from this that rhetoric may be generally defined as 'sign behaviour exhibiting a pragmatic concentration of meaning.'

In relation to this it may be noted that the map can be said to operate with all three discursive modes. It certainly implies 'factual reportage' as an intention, because it has a formal structure which controls the placing of signs within it, and therefore their location. In addition, some types of sign are used to indicate position, which could be conceived as giving it a syntactic structure. Because it is obviously intended to be useful to other people, it must have a pragmatic dimension. McNally

(1970: 76) makes the point that 'According to this view, although every instance of language exhibits all three kinds of discourse, one may also speak profitably of certain instances of language as being oriented, weighted or concentrated predominantly towards a semantic, syntactic or pragmatic dimension of meaning.'

Examples of all three 'concentrations' are not difficult to find in different types of map. The apparently impersonal topographic narrative of the national survey organization, with its emphasis on accuracy, standards and so on, is clearly semantically inclined. The nautical chart, with its employment of a specific geometrical structure, underlines the formal relationships between elements within it and is therefore syntactically oriented; and the statistical map illustration is intentionally the product of communication with a reader and therefore pragmatically oriented. Yet all of these must possess the three dimensions in varying degrees.

Meaning and interpretation

In discussing *Interpretation in Science and in Art*, Osborne (1986: 3–7) examines the connections between interpretation and meaning. Although, as he says, in a general sense 'a person who offers an interpretation of anything tacitly claims correctness for it,' it is evident that whether or not any interpretation of a map is 'correct' is often not verifiable, and indeed the map user may have no way of knowing whether or not the interpretation is correct in this sense. Osborne then makes the point that 'Interpretations of works of art are valued, not so much for correctness, as for their validity and perspicacity.' This lies much closer to many kinds of map interpretation, in which the realization of the connections between certain objects in space is perspicacious, and relevant to the map user. Osborne goes on to consider the kinds of communication which allow a multiplicity of meanings. In contrasting art and science, multiple meanings are what science tries to avoid, because ambiguity is a fault in factual communication. He also explains that 'explicitness must not be confused with superficiality or ease of apprehension. Meaning may be profound or difficult of access without being ambiguous.' This is relevant to the declaration sometimes made that all maps should be 'easy to read'. They can be legible yet still require effort and concentration by the map user. In contrasting the scientific concept of meaning with the artistic, Osborne maintains that 'The assumption that the object to be interpreted has or should have a single meaning in relation to which the interpretation is correct or incorrect no longer holds good in these contexts,' and therefore artistic works 'contain a multiplicity of valid but not always completely compatible meanings which become apparent only in the course of time and not all of which were consciously known to the artist or intended by him.' Without at this moment pursuing in detail the nature of the map as an artistic work, one may contrast the lucidity of these comments (and their relation to map interpretation) with the verbal entanglements of discourses on linguistic analysis.

The same principles can be applied to consideration of cartographic design, in the sense that all maps are themselves interpretations of phenomena based on information about them. Osborne (1986: 7) states that 'artists are said to "interpret" when they impose a coherent *perceptual* order upon a segment of perceived or imagined actuality or when they bring such a segment within a *perceptually* ordered structure.' On the other hand, 'scientists are said to "interpret" when they bring a section of experienced facts or observations within the ambit of a coherent *intellectual* structure, usually mathematically expressed.' If these two statements are taken together, they reflect the scientific and artistic components of maps. The structure of the two-dimensional plane as a representation of all or part of the Earth's surface, described by coordinate systems, must be intellectually coherent, although it is also iconic; whereas the symbolic graphic representation of objects or phenomena depends upon a coherent perceptual order which is fundamentally artistic.

These observations also yield an interesting sidelight on the frequently repeated call to make cartography more scientific – usually in relation to design, or more correctly, representation. In reality, the province of science in cartography is essentially concerned with the quality of locational information on which maps are based, not that of trying to establish rules for the unambiguous communication of messages. Indeed, if the aim of making map symbols correspond with meaning in a literal sense could be realized, it would at the same time destroy the value of the map as a means through which many interpretations can be conceived.

All artistic works are rich in possibilities. This underlines a fundamental difference in view between those who see the map as a communication device limited to the information intended in the input, and those who see a map as a representation, through which many interpretations are possible.

The power of map symbolism, like all graphics, is that of diversity in creation and combination. Language and graphic symbols operate in fundamentally different ways, and despite some brave attempts, there is no indication that the analysis of one produces an enlightened understanding of the other. Given this, one may ask the (rhetorical) question – why is it that there is such a concentration on the presumed analogy between map and language, and so little inquiry into the relationship between map and art?

CHAPTER 13

Cartographic theory and cartographic practice

Looking at the present rate of publication, in journals and books, on cartographic matters, it would seem that cartography is in a remarkably healthy state. Conferences, meetings, commissions, study groups – they all reinforce the impression of stimulating activity over a broad field of study. Although this may now be taken for granted, in fact it is a relatively recent phenomenon.

During the last fifty years, maps and map making have gone through many interesting changes and developments. Although the most obvious of these have been of a technical nature, publications about cartography and other map-making fields have developed to a degree that would have been regarded as entirely unlikely half a century ago. The current state of cartographic activity, in its broadest sense, is the product of many different factors, and although in a way this is well known (as it has all happened within a working lifetime), it is worth examining the general development in order to put recent discussions into perspective.

The initial impetus came after the Second World War, which itself had considerable influence on cartographic methods and operations (as wars often do). Up to this time very little attention had been given to cartography as a subject of study, especially in the English-speaking world. For surveying organizations, particularly national surveys, cartography was essentially map production, because of the need to draw all the map originals. The highly skilled craft of the cartographic draftsman was developed, in the normal industrial way, by apprentice-ship and training, within the employing organization. The draftsmen (and it was another of the male-dominated crafts) produced the entire map image in pen and ink (or even by engraving) including lettering and patterns.

In university departments of geography (most which were relatively recent additions to universities) some elements of map illustration were taught to geography students, usually under the heading of 'cartographic techiques', which sometimes included the study of projections as a separate subject. The concentration was on manuscript maps to illustrate geographical accounts. Given that this type of complete pen-and-ink drawing was difficult even for the skilled cartographic draftsman, it is not surprising that it tended to make little impression on students, unless they had a measure of graphic ability.

In the English-speaking world, there was a clear distinction between surveying and cartography. The notion of the topographer, the surveyor who also produced at least the initial version of the topographic map, had virtually disappeared, although it did continue at least in the title of many topographic surveyors in European countries. Commercial map and atlas producers, who did not produce the original source material, normally employed map editors or cartographic editors to devise, compile and design atlases and maps, but they were usually distinguished from the 'cartographers' who did the actual production.

At this time there were no organized meetings between cartographers or even cartographic editors. There were no journals on cartography in the English language. In Britain there were no public educational courses in any aspect of cartography. This, of course, was true of many manufacturing industries which had evolved from a variety of crafts. To those entering cartography today, the mental isolation which characterized the field must be hard to imagine.

Technological change

As is often the case, wartime pressure to produce maps rapidly not only led to changes in practice, but also brought many people into map production from other occupations. In the United States in particular, the need to produce maps more quickly led to the introduction, at least experimentally, of alternative production methods, such as scribing and typeset lettering. Although these were not entirely new ideas, they were given a new impetus. At the beginning of the 1950s the enterprising manager of a group of Swedish cartographic companies, Carl Mannerfelt, who had become aware of these new methods, was busy introducing them into commercial work.

Given that most European countries were faced with the mammoth task of replacing, recreating and developing maps and atlases of all kinds for the post-war period, it is not surprising that any new and faster production methods excited a great deal of interest. Thus began the first major technological transition of this half century, the graphic arts revolution. With the rapid development of the graphic arts in general, from printing to devising entirely new reprographic materials and processes, cartographic production moved steadily from a complete reliance on drawing in ink to a complex combination of manual and reprographic processes.

In all organizations where current production has to be maintained, making changes affects not only the actual methods used, but the organization and development of staff. New procedures have to be examined to see if they actually bring the supposed benefits (faster or more economical production), and this can only take place gradually. Staff have to make the transition to new methods, and where skill is involved, such radical changes can meet resistance. This transition to another technology has been repeated with the gradual development of digital mapping, made more complicated by the fact that as the change involves considerable capital investment, eventually this additional cost has to be recovered from the anticipated improvements.

Nevertheless the culmination of the changes brought about by the graphic arts revolution resulted in a remarkable increase in the quality of map production, and made a more refined use of the graphic variables possible. For example, a progressive scale of area colours depended on the ability to combine subtle variations in solids and tints in several colours, which was made possible by the new percentage screen tints. The effect of these changes, not only on the technology of production, but on the range and sophistication of cartographic representation, is demonstrated by maps and atlases of all kinds produced especially in the 1960s and 1970s, which set the standards which still characterize cartographic practice.

Cartographic organizations

Perhaps because of the more openly international attitude of Swedish society, Mannerfelt decided to organize an international conference of representatives from the cartographic community, both government and commercial, to discuss technical and other developments, and this took place at Tollare, near Stockholm, in 1956. Delegates were personally invited. All the papers were contributed by people working in some area of cartographic production or management. The initial proposal to consider setting up an international cartographic association was made at this meeting, and a small committee under Mannerfelt's chairmanship was organized to make further enquiries. Two years later Rand McNally hosted a similar meeting in Chicago. In 1960, the International Cartographic Association came into being.

Many problems were raised. Some (mainly European) countries already had representative cartographic societies, but others (such as Britain) had no mechanism for consulting 'cartographers' except through the management of governmental departments or private companies. The relationship between the proposed association and other associations in the map-making field had to be clarified. Whether participation should be by individuals or by organizations brought out very different responses. Although it was recognized that a few 'academic' cartographers might be interested, there was no suggestion that the association, if formed, should be an 'academic' organization.

Inevitably mistakes were made. Looking for an existing international organization which could act as an 'umbrella' for the fledgling cartographic association, Mannerfelt turned to the International Geographical Union. It was not a satisfactory relationship, and some years later had to be abandoned. But the movement did lead to other developments, in particular the founding of the British Cartographic Society, and a national committee for cartography in France. In the United States, it gave cartography a greater impetus within the

American Congress on Surveying and Mapping, and eventually the American Cartographic Association developed as a member organization. What was clear, despite variations in terminology, was that 'cartography' was used in its specific sense, and was distinguished from surveying and geodesy. This did not mean that those concerned were unaware of the role of surveying in map making. In fact, the foundation of a cartographic association made clear that the term 'cartography' was not being used to subsume the entire map-making field.

Cartographic publications

It was within this initial framework that publications in the English language specifically devoted to cartographic matters began to develop. In an interesting review and analysis, Gilmartin (1992) refers to the *Cartographic Journal* as an 'academic' journal, but this was not the intention when it was founded. Indeed, in the first issue, of the six main contributions, only one was from an 'academic'. Yet recent issues of the same journal are overwhelmingly dominated by authors holding academic positions. It is clear that within this period, writing about cartographic matters of all kinds has developed at a remarkable rate.

With the gradual establishment of educational courses in cartography, not only at university level but also in technical education, it was inevitable that a variety of textbooks should also appear. The idea of researching many aspects of cartography and map use steadily developed, and this in turn led to questions about the nature of cartography, its relationship with other fields, and the ideas and conceptions which underpinned cartographic practice. Inevitably, much of this development, certainly as far as publications were concerned, was driven by those who were interested in maps and cartography, but were not themselves practising cartographers. Indeed, very few practising cartographers made the transition into academic positions.

Not surprisingly, events and movements which influenced academic staff in colleges and universities tended to be imported into cartographic research. Given the need for university staff to do research and publish, societies and journals which initially were intended to be forums for practising cartographers were largely taken over by 'cartographers' whose entire working lives were spent in academic institutions. Although largely unrecognized, this development has had a remarkable effect on how cartographic matters have been treated, because so often any discussion of cartography tends to be considered in the light of what has been published about it.

Cartography and cartographers

The meaning of the term 'cartography' has been further confused because of a tendency for historians of cartography, especially in the United States, to use cartography as a general term for anything to do with maps and map making. This not only confuses them, it also confuses others. This also tends to be followed by some international organizations, so that the United Nations publication *World Cartography* includes material on geodesy and surveying as well as cartography proper. It is also reinforced by the particular needs and practices of some map authors, especially geographers. As they provide the initiation and the special information for many illustrative maps, they naturally tend to see 'cartography' as the whole map-making operation.

A similar degree of confusion arises over the use of the term 'cartographer', so much so that it has now become desirable to distinguish between practising cartographers (those who are employed in making maps) and 'cartographers' who are actually employed by other kinds of institution. Unfortunately. the distinction is rarely made, although Petchenik (1983: 37) drew attention to it in saying that 'specialization in cartography has developed to the point where academic studies of map design and map use may be completely divorced from the non-academic, routine map production milieu.'

Because the publications on cartography and map use are dominated by non-practising 'cartographers', their views and opinions tend to be the ones which are quoted, and which are available to students through books and journals. For example, Morrison (1976: 84), writing about the 'paradigm of cartography as a science of the communication of information,' could make the statement that 'Cartographic scientists in many nations are now accepting this paradigm,' although the 'cartographers' he referred to were in fact a small academic group which attended the same meetings and published in the same journals. The thousands of practising cartographers were never consulted of course. The effect is cumulative, for in some recent literature many basic considerations about cartography are drawn from what has been written, rather than what practising cartographers actually do. It remains a difficulty in any field in which the idea of 'knowledge' assumes that it must be embodied in language. Hence the near invisibility of cartographic 'art', which does not respond easily to verbal description. The same influence lies behind the constant meddling with 'definitions' of cartography, the effect of which is to reduce it to some amorphous verbal construction. Despite the great changes in technology, what cartographers actually do, and the nature of the problems which cartographers face, remain very much the same.

External influences on cartographic theory

The development of publications on cartographic matters covers a wide range. If the English-language journals are examined, together with many reports from both foreign and international meetings, it is clear that technology, in both theory and practice, has always received attention. Given that it affects the daily lives and problems of practising cartographers, this is natural and will continue. Articles on the history of cartography, together with papers on research into map use, have also provided a small but steady contribution. Apart from papers and even books on such specific areas, there has also been a fluctuating emphasis on what may be called general cartographic theory, in which the nature of cartography as a whole is examined, or some fundamental theory advanced. Significantly, virtually all such theorizing has been based on the introduction of ideas current in non-cartographic fields. Although the subject matter is often limited, it is used to make observations which are applied to cartography and cartographers as a whole.

Up to now, one common element in a series of different movements has been the constant emphasis on making cartography 'more scientific'. Psychophysical testing has been deliberately aimed at finding rules for the construction of proportional symbols on an experimental basis, following the practices of scientific investigation. The early forms of computer mapping, focusing on small-format black-and-white statistical maps, were themselves early attempts to generate certain types of map by computer processing. At least it was implied that using a computer was more 'scientific' than relying on the cartographic draftsman. In the early days of remote sensing, it was often suggested that the 'objective' data gathered by such means would make 'subjective' cartography of little importance, even to the extent of forecasting the demise of cartography as such.

The introduction of theoretical models on the part of the geographers reflected the ideas which were current in geography at the time. Board (1981: 51) stated quite plainly that 'For those who were connected with the exciting movements in geography, the atmosphere was just right and to an extent the temptation to emulate and imitate other geographers must have been significant.' The addition of information theory, as a component in the development of ideas of cartographic communication, was another reflection of external influence. Its effect can be seen in the terminology adopted, and to some extent still used, with its references to messages, coding and receivers. Converting cartography into a 'scientific discipline' (as the titles of many articles show) was an avowed objective of this general theorizing about maps and map use, to the point where any objection to it was made to appear to be merely the reluctance of 'traditional' cartographers to move into a proper scientific world. On the other hand, the real scientific aspects of map making, the use of coordinate systems to define location and the need for measurement in surveying and photogrammetry, were largely ignored.

It was during this particular evolution that the gap between what was being described as cartography, and the assumed attitudes of practising cartographers, became most marked. When cartography was being criticized, it was at least implied that the fault lay with practising cartographers. Even though the transition to digital technology – the second great technical change in this half century – was obviously going to take a long time for cartographic production organizations to digest, evaluate, and put into practice, they were constantly being urged to follow the new technology.

The ultimate model of this concept of scientific progress is essentially that in which the digital production of maps as part of geographic information systems is to be carried out entirely by digital computers, so that the cartographic tasks of composition, representation, design and production are executed by intelligent programs or by knowledge-based expert systems. These current areas of development will be examined later, but for the moment it can be accepted that in their initial form they also reinforce the idea of cartography as a scientific discipline with its own body of knowledge and experimentally tested rules. This knowledge, of course, had to be embodied in writing.

Now it can be argued that as practising cartographers had shown little inclination to develop the theoretical basis of their subject, it was right and proper that others should take on the task. But there are obvious reasons why cartographers appeared to contribute so little to the increasing intellectual debate. Petchenik (1985: 8) describes it succinctly:

> Practising cartographers tend to be so busy earning their livings by making and selling maps that there isn't 'free' time or energy left to be expended on research and writing projects: as a consequence, their point of view is not accurately reflected in the literature. The ultimate reward for mapping activity in the private sector is a profitable operation, or in the public sector, meeting institutional budgets and schedules, not in publishing articles or attending conferences.

In the new intellectual climate, practising cartographers were often accused of being too preoccupied with technology, especially by those who had no concern with map production. But this preoccupation is inevitable. Technological change has always affected cartographic practice, and will continue to do so. In any field which is concerned with making things, the means of making them is of vital concern. Map making has always borrowed from technical developments in other areas, and normally needs to modify and develop them for specific purposes. Cartographic digitizing is an obvious case. But the point is that research in these areas is driven by cartographic needs. Theorizing about cartography as a 'science' is not. It was essentially the result of attempts to bring the discussion of cartography (or what was assumed to be cartography) into terms which were regarded as intellectually acceptable within certain academic circles. The fact is that these intellectual shifts essentially follow fashions. For example, Taylor (1983: 6) could report that 'It was this feeling of unease with what was seen as a "non-rational" and "non-scientific" approach to map design which helped to stimulate the development of the theory of communication as a scientific basis for cartography.' Yet a decade later, the same author, Taylor (1993: 47), could comment that 'It has been realised that both the quantitative revolution and the communication model were in themselves sterile and certainly inadequate paradigms for geography and cartography respectively.'

Maps and social theory

The latest revelation to be imported into discussions of cartography is another product of academic influence. Partly a revolt against what is seen as the scientific bias of modern culture, and partly an attack on what its advocates regard as a pretence to neutrality and objectivity of the previously accepted 'establishment' views of history, literature, anthropology, etc., the chief line of the deconstructionist argument is that all communications and representations are loaded with intentions which underlie the ostensive meaning; and that therefore the truth can only be reached by deconstructing this image to find what lies behind it. In this respect, what any treatise says, in terms of its literal interpretation, is suspect. As all communication uses sign systems, then signification is not only in the literal meaning of the signs, but the way that things are presented and the ways in which certain things are included and others omitted. As a movement, it primarily affects those studies which operate in language – linguistics, literature, history, sociology, anthropology, etc. – and the area sometimes referred to as the 'human' sciences.

Not surprisingly, this provides a curious mixture. On the one hand, it contains some fundamental truths, which can be easily demonstrated by any examination of political image making in the Western world. But like other revolutions, it is easily taken to excess, so that one orthodoxy is replaced by another. Where it simply becomes another fashion, then it rapidly descends into a pseudo-intellectual movement which can be interpreted or applied in many different ways, and can be borrowed for many other purposes. Its chief interest cartographically is that now the claim has been made that cartography is 'too scientific', too much obsessed with notions of 'accuracy', and its practitioners too concerned with technology rather than the more important matters of the social relevance of maps, i.e. the way in which they portray 'the world'.

Despite the difference in origin and intention, this line of argument shares many of the characteristics of the theorizing on cartographic communication. It is extremely vague about any difference between cartography and map making. It involves assertions about cartography drawn entirely from what appears in books and journals. It constantly criticizes practising cartographers for what they have done and what they do. It apparently regards the map user as a passive receiver of messages, although now the 'message' is the function of the map to inculcate certain views of society. It is highly influenced by dealing with maps and cartography in language. And its arguments and propositions are not supported by examples drawn from a representative range of different types of maps. Indeed the gap between the high-sounding claims and the triviality of the few examples given is quite remarkable.

The reinterpretation of history

The rewriting of history happens in almost every generation, and produces interesting arguments among historians. The introduction of the concept of deconstructionism into cartography originated with the ideas of one historian of cartography, through which previous 'models' of historical writing were attacked. In itself, this is not unusual, as arguments about how the past should be interpreted are a necessary part of historical study. Whatever interpretation is made of the nature, function and significance of a map, the object itself remains the same, and can be reinterpreted by yet other scholars at some later date.

The principal material for this line of thought was presented in three papers by Harley, published in 1989, 1990 and 1991, together with some comments by other writers and a response by Harley (1989). Harley did not put forward any new cartographic theory as such, but nevertheless it is clear that he wanted to change the attitudes of cartographers and in some ways cartographic practice. Given that the articles by Harley are repetitive, frequently confusing and sometimes contradictory, it is not always easy to pin down the substance. This is unfortunate, because several serious issues are raised which are worth examination and discussion. Most of the 'arguments' are based on a whole series of unfounded assertions, and hardly anything is justified by evidence.

Linguistic jargon

Harley (1989: 3) begins by making clear that he intends to employ the ideas of deconstructionism, and therefore to 'read between the lines of the map.' Although his initial interest lay in the ways in which cartographic history was interpreted, this moves into a discussion of cartography as it appears to function in the present.

Just as communication brought with it a whole jargon of transmission, messages, information, etc., so the new mood can be readily identified by the constant repetition of certain words and phrases. These include power, hidden agenda, rhetoric, institutionalized power, ethics, text, etc. The arguments are expressed in dramatic terms, and there is presumed to be a sharp divide between those who will accept the new ideas and those who – it is assumed – will oppose them. In fact, the arguments are sustained by all sorts of unproven assertions, and the well-known 'rhetorical' device of misrepresenting the views of the 'opponent' in order to demolish them (the Socratic dialogue technique).

Some examples from Harley (1989: 3–14) will give the flavour: 'to locate the presence of power ... and its effects in all map knowledge'; the 'hidden agendas of cartography'; 'omnipresence of power in all knowledge'; 'search for metaphor and rhetoric in maps'; 'maps ... especially those produced and manipulated by the state'; 'Maps are still used to control our lives in innumerable ways'; 'Maps are authoritarian images'.

This language is used to make a series of claims, or assertions, about cartography, cartographers and maps. Some of the principal claims are as follows.

Cartography and science

In discussing the attitudes and supposed beliefs of cartographers, Harley inevitably confuses what cartographers do with what is written about cartography mainly by academics. He states (Harley, 1989: 2) that 'Many find it surprising that "art" no longer exists in "professional" cartography.' In fact, it is inevitably part of cartographic work, and exists whether or not it is written about. The notion that cartography should be, or is, entirely 'scientific' is imposed by those who espouse communication or other 'scientific' models.

He goes on to state that cartographic art 'has often been accorded a cosmetic rather than a central role in communication' (Harley, 1989: 4). This again is not the view of practising cartographers, but of those who write about cartographic communication theories, and frequently cartographic historians. Repeating the notion of the map as a 'mirror of nature', a view largely generated by art historians, Harley merely ascribes this to cartographers, without bothering to find any evidence to support it. Yet the question can be asked: does any practising cartographer believe that a map is simply a 'mirror of nature'? If, as Harley subsequently states (1990: 1), 'maps represent the world through a veil of ideology' how can they be a mirror of nature?

In the same paper, Harley (1990: 10), reflecting on events since the 1960s, observes that 'Cartography was to acquire the status of a sub-science.' This, of course, was the view and intention of theories of cartographic communication, not of the unconsulted 'cartographers'.

Another line of thought is that cartographers are over concerned with accuracy, with imparting the notion that the map is a neutral, objective representation, and that this reinforces the scientific view of the world which has been the dominant force in society and culture. Harley assumes that this is a deliberate act on the part of cartographers. What he fails either to understand or reveal is that far from being some kind of cartographic wish fulfilment, a concern with accuracy is inevitable because of the nature of the map. In dealing with location, a map, unlike language, cannot be vague, ambiguous or indefinite. Any symbol in a map is assigned a specific position, and it states this position without qualification. A map cannot say 'I think it may be over there'. It is this particular property, which does not exist in natural language, that makes maps useful. For whatever purposes the map is devised, and whether these are regarded as socially proper or malign, unless the map is 'accurate' it is meaningless.

Harley would probably have argued that even so cartographers have not been sufficiently concerned with the social role of maps, but this is an entirely different matter. Linking this with 'accuracy' is misleading. The socially 'correct' maps envisaged by Harley would still need to be 'accurate' if they were to serve their purpose.

The hidden agenda

A recurring theme in Harley's articles is that there are 'hidden agendas of cartography' and that 'Our task is to search for the social forces that have structured cartography and to locate the presence of power – and its effects – in all map knowledge' (Harley, 1989: 2). The fact that some historians of cartography have discovered that map making functions in a real and imperfect world is to be commended, but it is hardly surprising news for cartographers. Maps are made to be used, and as such they reflect the needs of the societies for which they are produced.

Anyone with a general knowledge of the history of map making can provide some obvious examples. Many of the national surveys in Europe were founded initially for military purposes, as their original names and staff control indicate. This is not exactly a revelation. In the 1820s, the Ordnance Survey was largely moved to Ireland, to carry out a new topographic survey, the aim of which was to make possible the description of the townlands for purposes of taxation. Even the scale was chosen for this purpose. This agenda was not 'hidden'; it was the stated reason. At the end of the eighteenth century, when the Admiralty began to grapple with the problems and possibilities of nautical charts, the basic reasons were quite clear. The Royal Navy needed charts to carry out its functions on behalf of the state, and the merchants were worried about the loss of ships and cargoes. Safe navigation in coastal waters also meant that seamen's lives would be saved, but this was not the main consideration, as they could easily be replaced. The English county maps produced during the eighteenth century were frequently embellished with the coats of arms of the aristocracy and the seats of the landed gentry. These were the principal patrons and subscribers who paid for the survey, and therefore provided the 'market'. The 'cartographers' (who were primarily surveyors) who worked in this field no doubt differed in their intentions. For many, making and selling maps was seen to be a profitable business; others had a genuine interest in map making and saw their work as having great social and practical utility.

The fact that map makers operate within the political and social structure of the societies in which they live does not have to be revealed by 'deconstructing' any myth. The many enquiries into the work and programme of the Ordnance Survey, from the middle of the nineteenth century to the present, were carried out by and on behalf of government, and those sections of society, such as local authorities, who were regarded as the chief users of the large-scale plans. Indeed the introduction of the large-scale (1 : 2500) plans by the Ordnance Survey was a direct reflection of the new administrative responsibilities placed on local authorities by many Acts of Parliament dealing with education, sanitation and public services. This is not 'hidden'; it is part of the published record. The commercial production of special maps for cyclists from the 1890s onwards was a result of the increased leisure of the 'working class', shorter working weeks and eventually holidays with pay. These social and economic changes were reflected in the maps that were produced to satisfy this market.

The question of how the production and publication of maps is controlled in any society is an interesting and important issue, but it is not illuminated by uttering cliches about hidden agendas. Nor does its examination require any deconstructionist philosophy. In commenting on the disproportionate emphasis on research into symbol design, Petchenik raised the issue in her usual lucid manner:

> Most maps used by most people are issued either by commercial firms or by government agencies. Yet academic cartographers have all but ignored the process by which these map makers decide what content will be mapped for widespread distribution. Insofar as shaping a culture's image of space is concerned, *these* are the truly fundamental decisions – far more important than decisions about type faces or layer-tints. We need a more accurate and comprehensive understanding of what 'drives' the map making and map using system in our culture. Why do we map, what do we map, and (ultimately) who pays for it? (Petchenik, 1983: 65).

The map as rhetoric

Although Harley (1989: 11) refers to classical rhetoric, it is clear that he still interprets rhetoric as essentially a means of persuading the 'audience' to accept something which is concealed, i.e. the true intention of the initiator. Once again he puts words into the mouths of 'cartographers'. He asserts that 'for some cartographers the notion of "rhetoric" would remain a pejorative term,' without bothering to quote any sources. The same sort of misrepresentation appears in his reference to 'The hysteria among leading cartographers at the popularity of the Peters projection' (Harley, 1989: 5). It was not the so-called popularity that was objected to, but the unfounded and unproven claim that the projection had some unique virtue (see Loxton, 1985).

The study of rhetoric stresses the relationship of content, structure and expression. Successful communication, for any ends, achieves an appropriate balance between them; unsuccessful communication does not. To apply the same criterion to Harley's writing, it is unsuccessful because it is unconvincing, despite the emotive power of language which he consistently employed. It is unconvincing because the propositions are not backed by evidence. If he had actually demonstrated, in detail, that 'cartography is an act of persuasive communication' (Harley, 1989: 11), then his arguments would carry more weight. This can only be the case if some map user or group of users is persuaded. But the range of possible responses – or lack of them – is simply ignored.

Most of the time the references to 'map' reflect some vague general notion that is essentially linguistic in nature. For Harley, all maps involve 'appeals to the potential readership through the use of colours, decoration, typography, dedications, or written justifications of their method' (Harley, 1989: 11). But no specific examples are described in detail to show how this takes place.

The map as a text

Not surprisingly, Harley's (1989) treatment of this idea is entirely rooted in language. Given that the philosophers and writers he quotes are all people whose activities are essentially verbal, who express themselves in language, it is hardly surprising that the linguistic model is accepted without question. He simply avoids any consideration of the complexity of the comparison by asserting that 'It is now generally accepted that the model of text can have a much wider application than to literary texts alone. To non-book texts such as musical compositions and architectural structures we can confidently add the graphic texts we call maps' (Harley, 1989: 7). (One can note the persuasive effect of the collective 'we' and 'confidently'.) It can be argued that maps as 'constructions' are fundamentally different from linguistic texts, but no such possibility is considered here. No one, for example, has been able to demonstrate that a map is 'read' like a linguistic text. If there is a useful analogy, then it needs discussion and detailed analysis, but that would uncover problems that are not 'appropriate' to this discourse.

The map as an argument

The term 'argument' can be interpreted in several ways, but it normally implies the existence of some other point of view. A proposition sustained by reason and evidence is advanced with the idea of convincing people to accept it in preference to some existing or presumed alternative. The concept is interesting. But if a map states an argument, and this argument is challenged as being false in some way, then the only appropriate way of pursuing it would be to create a different version of the map to represent the alternative argument. Given that much of Harley's criticism seems to be applied to topographic maps (it is rarely made clear what sort of map is under question), then the argument that a map should reflect social concerns could be suitably and convincingly illustrated by actually producing at least a specimen sheet of this proposed map.

To do this, of course, would require more than vague verbal advice, or adding a few more symbol categories. Actually producing such a map, and especially one with an original conception and design, would involve a considerable outlay of time, effort, skill and resources. If the academic critics would engage themselves in such an operation, thereby having to define exactly what areas are to be described, for example, as 'toxic waste sites' or 'deprived inner cities', and by what criteria, they might come to a better understanding of the problems of the cartography they claim to know so much about.

If indeed there is a market for maps which reflect the interests and beliefs of the socially concerned, in which, for example, due attention would be given to toxic waste sites, polluted rivers and derelict or deprived inner cities, then it could be tested by producing and publishing at least one or two examples. But actually making maps inevitably turns out to be something that the academic critics leave to others. And even if this could be achieved, presumably such maps would also have to be deconstructed by yet other critics.

Fact or myth?

It is revealing to apply Harley's contentions to a range of actual maps. Taking a very simple one to begin with, I frequently use a street guide of my nearest town. Recently I wanted to find the location of my dentist's address. The large-scale plan is published by a small local company, who presumably saw a prospective market for this kind of information, and hoped to make a modest profit. The plan indicates the streets and buildings. It is not a 'cadastral' plan and does not show property lines. It identifies and names various public buildings, post office, churches and so on, and other public facilities such as parks, car parks, the riverside walk, etc. It is produced in black and white, and does not carry any advertising, only the name of the publisher and printer.

Where is the 'hidden agenda'? What am I being persuaded to believe that is cloaked in secrecy? What 'institutionalized power' does the map reflect? Certainly the map gives me the power to find places I want to locate, but that depends on my requirements and interpretation. What 'ethical' statements does it make? What 'argument' is it proposing? In fact, this bare outline would be unintelligible to a person unfamiliar with an urban environment. What it means is largely a function of my comprehension and knowledge.

Of course, a great deal is omitted. It does not state whether or not the river is polluted; it does not tell me where poor people or rich people live; it does not describe social conditions; it does not tell me whether or not the streets are full of traffic. This is not its purpose. It does include verifiable facts about the location of places, and if it did not it would be useless. These facts are very properly the product of scientific survey. It operates within the culture and needs of present-day society, and makes no comments upon it. It makes possible a few of the innumerable ordinary uses which maps serve without claiming to reveal any of the complex forces of modern society. In all, it seems to have little to do with the pretentious language of deconstructionism.

For a second example, the coastal nautical chart is an important type within the general classification of map. Its information is based on the scientific work of both topographic and hydrographic surveyors. The classification of features, landmarks, navigation systems, aids and dangers is that required by navigators. The content is not chosen by 'cartographers', who are distinctly not allowed to omit whatever they wish. Even the treatment of the coastal topography is based on what can be observed from seaward. The colours used are highly conventional, and any changes to the symbol specifications are always resisted by mariners.

Of course, the charts can be used for naval defence. Of course, one function of such charts is to enable the safe transport of valuable cargoes which make profits. But as this is regarded as a central objective in 'Western' society, there is nothing strange about that. Is this what is meant by the 'hidden agenda'? If so, It is not very well hidden.

What 'distinctions of class and power' (Harley, 1989: 7) are to be found among the buoys and the sandbanks? What propositions do they promote? What 'argument about the world' (Harley, 1989: 11) is concealed between the sounded depths and the marked navigation channels? It is no use maintaining that of course Harley was referring to some other kind of map. Harley makes no such qualification, and therefore his observations must be applicable in all cases.

A geological map is an example of the type of special-subject map resulting from the application of science to the description and understanding of the physical world. The classifications shown are a function of geological science, not of cartography. The use of many variations of colour and pattern is to provide a legible visual order for a large number of area symbols. Harley (1989: 11) maintains that the use of colour, among other devices, 'appeals to a potential readership.' Having been involved in the cartographic production of more than one basic geological sheet, I can *confidently* state that any such appeal was far from my mind. I was much too preoccupied with the problems of contrast and visual order, i.e. the 'art' of cartography.

Presumably the 'hidden agenda' here applies to the fact that such maps can be used for the exploitation of minerals, and therefore the destruction of the environment. But those activities took place long before geology was studied as a science or any geological maps produced. What 'social consequences of cartographic practice' (Harley, 1989: 8) are hidden among these classifications of different types of rock? What 'myth' do geological map sheets project? Harley would no doubt have suggested that all such scientific investigation supports and reinforces the idea that society progresses by accumulating knowledge about the external world, but that is a challenge to society, not the process of map making.

One of the few examples that Harley (1989) quotes, with approval, is the well-known analysis of the State Highway Map of North Carolina (Wood and Fels, 1986). Apparently embedded in a package of non-cartographic publicity, the addition of a variety of emblems, messages, flags, etc., is a fair example of the well-known practice of advertising the product in the most favourable way. Given the idealization of 'their' State by American citizens, this is hardly unusual. Given the reaction expressed by Wood and Fels, as rhetoric it was positively unsuccessful.

There are many other kinds of road map, all commercially produced for profit, especially in Europe. Accepting the dominance of roads in the developed state, and the omnipresent motor vehicle, this is hardly surprising. Such maps are highly selective. The classified road symbols state not only a physical description (number of lanes, etc.), but are also appraisive in the sense of describing a class of road within a hierarchy. Their description of the landscape is minimal. They do not describe traffic, driving regulations. quickest journeys, etc., although maps of this kind have been attempted (Morrison, 1980). Depending on scale, they can be used both to plan journeys and also to check on navigation during journeys. Unless they are 'accurate' they are useless.

Again, where is the 'hidden agenda'? It can be argued that the existence of such maps reinforces the antisocial nature of motor vehicles, the increase in pollution, etc. But if all such maps were removed, would there be any less traffic? It would certainly result in a great waste of time and even more pollution. The 'world view' that promotes the consumption of resources at the expense of pollution of the environment is not going to be effectively challenged by rearranging road maps. If the map is indeed '...an ethical statement about the world' (Harley, 1990: 6), the problem is not the map but the 'ethics' of the societies which need to produce and use them.

For a final example, the content of world atlases tends to be selected from the point of view of a particular country, and therefore reflects the interests (and frequently the school curricula) of the inhabitants. For most people, 'my' country is here, and other countries are further away. Most British atlases show the British Isles in greatest detail, European countries next, other parts of the English-speaking world and the former empire, and finally the 'distant' regions like Eastern Asia and South America. There are many historical factors in this 'world view'. A school atlas of Nigeria (produced by a British publisher) has Nigeria at large scale, followed by West Africa, Africa, Europe and Western Asia, and then the rest of the world. Far from being part of the some great plot to pervert people's understanding, the maps are the product of the ideas that already exist. Maps of the world produced in Europe have Europe 'in the middle'. What else would anyone expect?

Although the cartographic characteristics which Harley described are presumed to influence map users, map users hardly appear at all in his articles. Yet it is quite clear that most people spend very little time looking at maps, and a large proportion of the population never does. A simple outline map about some news item may appear briefly on the television screen, but this does not prove that the viewers actually look at it. The notion that anything so complex as a 'world view' is fundamentally affected by maps is without foundation. For the ordinary citizen, using a map is usually a short-lived affair. A question arises – 'Which road do I take to get to so-and-so?' – the road atlas is found, present position located, route number identified, and the map abandoned. The notion that such users are likely to browse in fascination among the appealing colours and secret messages is pure fancy. They are not interested in looking in the margins, or at the blank areas; in many cases they do not even look at the scale. This may be what scholars do while researching in map collections, but it does not characterize ordinary map use. Even among those who use maps professionally, as part of their work or their research, the point of interest is not the map, but the information it contains or the purpose it serves. People use maps to get what *they* want, not to expose themselves to some sort of cartographic brainwashing.

Cartographers, maps and morals

Despite Harley's attacks on the way maps have been used, and despite the level of criticism of the maps that have been produced, many of his remarks remain ambiguous. He states (Harley, 1990: 18) that 'Maps should continue to reflect the values of society at large, as their history has shown us they have always done.' But surely, reflecting the values of society at large is precisely what he had been criticizing, and what he wanted to change. Society at large, how ever unfortunately, is still dominated by governmental and corporate business power, whatever lip-service is paid to 'participatory democracy. He may complain that these values are not the ones he would prefer, but they do reflect the power of the state and the market economy.

'Urban blight or toxic waste' (Harley, 1990) are given as examples of what Harley thought maps should include. But these are the symptoms, not the disease. The problems of modern 'Western' society require a more profound investigation than can be provided by exhortations to cartographers.

There is little doubt that Harley's comments on ethics and cartographic practice will be those which arouse most interest. But although Harley referred to the cartographic profession, it is doubtful if his real interest was in creating a talking shop about 'professional ethics'. There are already signs that this new fashion will receive a great deal of attention, before being replaced by another. In 1991 the International Cartographic Association set up a working group to define the main theoretical issues on cartography, as reported in *Cartographica* **30**(4) 1993. In the introduction to this report, in the section on social context, Kanakubo and Morita (1993: ix) state that 'Until recent years. the studies on map making have been somewhat traditional. Studies of the relationship between the map and human beings had already begun in the 1960s with the development of the theory of communication. The studies on map and society will be the next issue.' The blandness of this comment, even by academic standards, is remarkable. Given that the working group is the eventual successor to the original Commission on Cartographic Communication, a great deal of significance is omitted. The objective of communication theory was to make cartography more scientific; the objective of 'social context' is to turn attention away from a preoccupation with scientific theory and technology towards cartography 'as a social-historical activity' (Torok, 1993: 9). The fact that these attitudes are diametrically opposed goes unnoticed, or is accepted without comment.

The question of the role of ethics in cartography, or more directly, the relationship between cartographers and ethical practices, is most emphatically expressed by Harley (1991), although this paper in turn is full of obscure references and confusions between cartography and map making. For example, Harley (1991: 13) refers to the problem that 'the content of maps will increasingly become a moral dilemma for cartographers if they accept their responsibilities for reconstructing the world that the surveyor has deconstructed.' But the work of the surveyor is part of the 'institutionalised power' which Harley has been busy deconstructing. Is this a deconstruction of a deconstruction? Like some other historians, Harley seems always to think back to an idealized notion of an individually named cartographer, who could be treated like any other author, and who was presumed to be in entire control of the mapping operation.

The central question for Harley can be represented by the following statement (Harley, 1991: 14): 'What are the principles of social justice that ought to be endorsed by cartographers? Should maps merely be an inert mirror of majority values or can they play a wider role in the struggle for social improvement?' But a struggle for social improvement only takes real form by defining in specific terms what sort of society is needed to provide 'social justice'.

The 'struggle for social improvement' and the ethical values of any particular society are political matters, not cartographic ones. Constructing some high-sounding description of 'ethical standards' in cartographic practice, or some sort of 'professional code', is not what Harley had in mind. What Harley wanted, such as the proper respect for ethnic minorities, the removal of racially offensive place names, and so on, can only be brought about by political action. It was the political emancipation of black Americans which brought about the recognition that such names are offensive; changing mapping policies was a subsequent event. But this issue is avoided. In that Harley was largely drawing on examples from American society, the dilemma he refers to, but only indirectly, is how individuals who wish to have a socially just society can achieve it when the society of the nation state they live in, and which fundamentally they do not challenge, is founded on the institutionalized power of corporate business and the 'ethics' of the market economy. To try to change this by altering the views of cartographers is of course political criticism from a safe distance. It confuses the message with the messenger.

Control of information, control of the media, secrecy in government, surveillance of individuals, and suppression of information are all factors which affect maps and geographical information systems. Harley goes so far as suggesting that: 'Indeed, the final ethical question may be one of just how far cartographers of all shades of opinion are prepared to be politically active in altering the conditions under which they make their maps' (Harley, 1991: 14). But at this level, cartographers are citizens, and have the same duties and obligations as other citizens. The fact is that Harley's view of 'social justice' is obviously not shared by the majority of American citizens, and indeed one dilemma for American society (and Western society as a whole) is that ethically it is deeply divided. As many writers have pointed out, one reason why so many special groups – religious, racial, feminist, etc. – have come into existence is because people find themselves living in a society which fundamentally recognizes no goal except profit and offers no aspiration to the sense of community they feel they need.

Given that maps can only be changed by the people who actually control the resources for making them, the fate of any cartographer who actively challenged the basic policies of the state and its official agencies, or the practices of commercial business, would be obvious. How many academics, quite content to advise practising cartographers about their 'ethical' behaviour, would resign their well-paid posts in protest about a matter of political principle?

As Salkie (1992: 183) points out, in Chomsky's view, 'Access to promotion and publishers, sponsorship, contacts wth influential people, and lucrative research contracts, all tend to go along with conformity. People who raise questions of fundamental principle are often seen as an embarrassment.' Chomsky is perfectly clear about the distinction between his research field and his duty as a citizen. But he is prepared to sacrifice his career for political principle. He is also quite open about the nature of the society he wants, which is in total contrast to the technology-driven, consumerist, ruthless 'market' economy and the unholy alliance of corporate business and state power. If questioning this in principle is what 'ethics in cartography' is about, then it deserves support. But a scrutiny of what has been discussed under the heading of cartographic theory in the last thirty years suggests that this interest in ethics and society will be no more than another short-lived fashion.

The theoretical problem

This brief review of some of the trends in thinking about cartography indicates that most of the proposals advanced at different times have originated in movements from outside map making itself, and have reflected the intellectual fashions of the time. Because of this their initiators seem unable to grasp a central issue, which is that cartography is both an applied science and an applied art. Unfortunately, these two dimensions are invariably treated as being in opposition. Any real theory of map making, even if it is only devised for discussion, rather than practice, has to incorporate these two elements, and show that a successful map depends on both of them being treated appropriately. Cartographers move from plotting a coordinate position to thinking out a suitable pattern for an area symbol, and they do so without separating one sort of knowledge from another. It is this combination of science and art which makes cartography interesting, and the study and use of maps so rewarding. If those who enjoy theorizing about cartography would start from this point, then some real progress might be made.

By drawing attention to the content of maps, Harley reinforced the importance of the appropriateness of the map content, as well as the quality, reliability, consistency and completeness of the information, within its limits of scale and generalization. This is just as important to the map user as the appearance and design of the map. The control of the resources needed for the production and publication of maps, and the political implications of making map information available, will always have an effect on which maps are made. In this respect, if the interests of map users are being considered, it is impossible to devise any useful general theory of map making by dealing with cartography in isolation, or even treating cartographic design and representation as the key element in satisfying the map users' needs.

The relationship between the 'real' world, obtaining mappable information about it, and the needs, interests and perceptual skills of map users cannot be described satisfactorily by reference to 'messages' or 'cognitive realms'. Nor can any deconstructionist philosophy be employed to renew cartographic practice, because how ever successful it may appear to be in breaking things down, it offers no guidance on actually creating them. If any general theory is to be devised, then the nature and limitations of the source material are a vital element in the whole map-making–map-using process.

PART FOUR

The map as an artistic work

CHAPTER 14

Artistic works and works of art

The preceding chapters have advanced the proposition that since maps depend on graphic means of expression, as presentational forms they are closely related to other products of the graphic arts. This inevitably leads to the further question of whether any map can be regarded as an artistic work, and if so in what way.

Before examining this, it is necessary to restate that the general term 'map' covers a great variety of individual objects, which differ considerably both in content and intention. A cadastral plan of delimited plots of land, and an aeronautical chart are both maps, but they differ in content, purpose and elaboration of representation. One is graphically simple, the other graphically complex. For artistic works this is not unusual, for a line caricature and a landscape painting indicate a similar range. Indeed this possibility of range is one of the characteristics of art.

The question of what distinguishes artistic works from other artifacts is one which has been argued at length under many headings. It has been examined by considering the distinctions between art and science; between objective and subjective; between poetry and descriptive prose; between functional and aesthetic properties, and there is a wide range of opinions. It is the particular concern of those who study aesthetics, because it is accepted that all works of art have aesthetic properties, however difficult it may be to define them. The frequent references to artistic skill in cartography suggest that some such consideration is also present, but there is rarely any attempt to define it in more specific terms.

Applied science and applied art

A map is an artifact, and requires the application of both science and art in its construction. Even the most graphically simple maps, containing the minimum of specific reference to the external world, depend at some point on information that has been produced by measurement. In addition, much of the subject matter of specialized maps is the result of

207

scientific inquiry in fields that are as wide as human knowledge of the physical and social environment permits. The modern topographic map, on which so many others are based, is fundamentally a product of scientific development and its application to map making. The fact that much of this development came about in pursuit of other ends is also characteristic of the map-making process.

Nevertheless, the representation of information in map form is graphic, and therefore must lead to consideration of the map as an artistic work. If a map is essentially artistic in form, then this must hold regardless of the content, or the technology used in its production. In this sense, if the artistic property of any map was simply dependent on the manual skill in drawing which for so long supported its production, it would no longer exist if this skill was replaced by a machine.

It is clear that as a map is intended to be useful, and as this is the primary aim in its production, it differs from those works of 'pure' or 'fine' art that do not embody any such intention. On the other hand, as an artifact the map is only one among many where the visible form is inextricably bound up with the function itself. Whether or not functional tools can also be regarded as artistic is therefore highly significant.

The normal distinction between fine art and applied art is also critical to any discussion of maps. If the term 'art', and therefore the possibility of aesthetic property, is limited to those things in which the aesthetic property is the only thing of importance, this would seem to exclude all other objects which are created to serve some other purpose. This in turn leads to a view that for such objects art is limited to decoration or ornament. As Berleant (1970: 9) expresses it, 'Another assumption is the conviction that there is a qualitative difference between fine art and the popular or practical arts, and that the same aesthetic criteria do not apply.' The narrow view is challenged by many philosophers. Wolterstorff (1980: xi) insists that 'the expression of emotion is but one among other reasons for the creation of works of art, and in the total body of the world's art, a thoroughly subordinate reason.' And he continues, 'In many, if not most cases, it was never intended by the artist that his work should serve as an object of aesthetic contemplation.'

The same point is made by Mukařovský (1978: xxxiv) in saying that 'one and the same object can be evaluated both as an implement and as a work of art.' Langer (1957: 121) takes the widest view, insisting that 'throughout the ages, the practice of artistic creation has seized on whatever occasions life offered; and artistic excellence has usually been felt as an enhancement of whatever other excellence the constructed object had.' This underlines the difficulty that critics have (in the analytical sense) of distinguishing between artistic and non-artistic values. It also begins to explain how it is that the expressive powers of art can be exploited to advance non-functional and non-aesthetic intentions, as is the case with so much modern advertising.

This dual characterization of a map as both a functional object and an artistic work has parallels in many other applied arts and sciences.

Architecture, with its combination of functional purpose and visual quality, is an obvious case in point. The analogy has often been invoked, because for a long period important buildings were conceived and executed by master builders, before the separation between design and construction characteristic of the modern era. In map production the combination of design and execution (with its inevitable reliance on convention and practice) which characterized the long period of both manuscript and engraved maps was eventually replaced by a similar separation.

In architecture, as with maps, a great variety of products is normal. Many buildings, conforming to current norms of style and function, are little more than repetitions of a familiar design. Major works, offered to open competition, embody much more directly the intention of original artistic work, in which aesthetic properties affect choice. At the same time, design is given a position, a degree of importance quite uncommon in the world of map making. Has any new national topographic map series ever been offered to competitive designs, with suitable financial incentives? Yet the total commitment of resources, in financial terms, is often not so dissimilar.

This visual quality in architecture combines both major and minor forms. Not only are the principal dimensions and shapes vitally important but the texture and decoration of surfaces must reinforce a unifying theme. In architecture, at least, there is a concept of designing the whole which is notably lacking in a great deal of map production, where design is often conceived as no more than the specification of individual symbols. The limitations of this approach are evident in the numbers of inadequate maps. This inadequacy stems not so much from a lack of resources, as the inability to apply the resources so that they act in combination.

Despite this, the analogy has to be used with discretion. There is a fundamental difference between a building and a map. The building itself is the function; it does not represent something else, and therefore the product of the design is what is used. But a map provides information about something which is external to it; what it offers is useful, not the map as an object itself. In this respect it is closer to a painting. Despite the fact that a painting has to be embodied (it has to be concretely expressed in a medium), it is not the painting itself but what it refers to or expresses that is important.

In this respect maps also have analogies with photographs, although with photography there tends to be a more distinct separation between its application to informative purposes, and its application to produce artistic works. Nevertheless, the best examples of photographic journalism, or even advertising photography, provide images that are both expressive and have aesthetic value. The truly creative photograph arrests the observer and demands attention. It is also the case that often the quality of the photograph as an art work is not dependent on sophisticated technology. Many of the greatest examples of photographic art have been produced with simple apparatus. It is what lies in the observant eye of the photographer which creates the quality.

Much of the emotive use of photography in advertising provides an excellent example of rhetoric in its pejorative sense. The images are designed to arouse associations of health, or prosperity, or success, in the mind of the recipient, and to attach these to the product being advertised. This is the deliberate abuse of aesthetic properties to persuade, to serve an ulterior motive, and it depends on creating an illusion intended to affect the imagination.

Consideration of this problem can illuminate several factors which are relevant to the question of whether a map can be said to be artistic, and whether this is simply a superficial addition, or a fundamental property. To pursue this it is necessary to discuss the relationship between aesthetic properties and design, the role of creativity and imagination, the relevance of intention in the making of art works, and the role of the perceiver.

Form and function

The notion of efficiency in the communication of information can be taken to suggest that any informational device should be functional – that its design should be fully controlled by what is necessary to its purpose. The same line of argument is often used to support the idea that the designer should therefore concentrate on satisfying the requirements of the user, without being distracted by any concern with superficial appearance or 'art'.

In a stimulating examination of design, Pye reveals the complete inadequacy of the notions commonly associated with the idea that function can be used as a controlling factor in design. Function, he points out, is not an isolatable property. Everything that is designed to satisfy some purpose, to achieve some end, takes with it a set of additional and unwanted elements. There may be, and usually are, alternative choices, but they are not wholly avoidable. In Pye's words 'It is quite impossible for any design to be the "logical outcome of the requirements" simply because, the requirements being in conflict, their logical outcome is an impossibility' (Pye, 1978: 70).

The point can be demonstrated by considering the problem of sheet size in a printed map series covering a large area. Ease of handling suggests that the sheets should be small. Continuity of information and economy of production suggest that they should be as large as possible. Relatively large sheets are the obvious preference of the topographic map production manager, although they may not be very popular with the people who work in production. They are anathema to the librarian and the shopkeeper who have to store and handle them in large numbers. A map series which is intended to be portable, such as a road atlas, can be made small enough for easy carriage and handling, but then breaks up the information into a collection of bits, involving frequent searches for the required map pages along a route. Overlaps between pages make this easier, but they also 'waste' space and make the product more expensive.

Printing the maps on paper has many advantages. It is widely available and relatively inexpensive. It can be printed on, cut and folded with comparative ease. It also tears easily, gets dirty, and is easily damaged if wetted. Alternatives exist, such as synthetic paper, but these are more difficult to print on and more expensive. As Pye (1978: 34) says, 'all useful devices have got to do useless things which no one wants them to do.' Design has to manage the conflicting demands which are imposed on any product. The map is no exception.

Provided that the product is usable in the minimum sense, and its appearance does not depart radically from the expected, concern with function is more commonly a matter of economy and ease of production. According to Pye (1978: 34), 'It seems to be invariably true that those characteristics which lead people to call a design functional are derived from the requirements of economy and not of use.... Streamlining, omission of ornament, exposure of structural members, "stark simplicity", all of these derive directly or indirectly from requirements of economy.'

The appearance of an object is never ignored, even though, as Pye points out, it often involves work which cannot be called directly functional. On many maps, whether the parallel fine lines representing a major road are exactly at the specified dimensions and exactly parallel is not of any real consequence. It would not have any noticeable effect on the usefulness of the map if they were imperfect. But few producers (or users) would regard imperfect lines as adequate. Clean printed tints, exact colour match between sheets, dots which are perfectly circular, even strokes in hachures, parallel lines in patterns – none of these things are essential in the functional sense, but they matter considerably to the user.

Aesthetic theories

If works of art are regarded as possessing aesthetic properties, then it would seem that such properties should be capable of definition, or at least approximate description. The term 'aesthetic' is strictly defined as pertaining to 'Things perceptible by the senses' (*Shorter Oxford English Dictionary*), and therefore is associated with what we normally regard as feeling, and the condition of being emotionally affected. In this respect, also, it is an important aspect of rhetoric, which treats what can be emotionally affective as a necessary component of communication.

In proposing an aesthetic theory of art, Eldridge (1985: 308) states that 'The aesthetic quality possession of which is necessary and sufficient for a thing's being art is the satisfying appropriateness to one another of a thing's form and content.' He regards this as the minimum definition, allowing room for new contents and new forms. It does not exclude any object on the grounds of whether or not it is classified as 'art'. As he says (Eldridge, 1985: 309) there are no 'rules for artistic creation.... Nor can what will be successful be predicted.' He argues that therefore 'the theory of art is both affiliated with psychology and not reducible to empirical laboratory psychology' (Eldridge, 1985: 311). This is so because a work of art engages the sensibility, which is quite different to acting predictably upon a passive receiver.

What engages the sensibility is that beauty can be found in all manner of things which are made for some purpose, not only those made as 'art'. Pye (1978: 96) makes the point plainly: 'The most unassuming, unpretentious and unemphatic things may be beautiful. Anything in the world may be beautiful and an endless multitude of things in it are beautiful: they "look right".'

Many writers, such as Langer (1957), Scruton (1974), Wolterstorff (1980) and Berleant (1970), refer to elements such as expression, imitation, intention, imagination, immediacy, metaphor, communication, and the uniqueness of any art work. Many of these need to be examined in order to see if they apply to maps and map making, and if so, in what way.

Expression

The terms expression, expressive and expressiveness occur frequently in aesthetic accounts. Although the simple denotation of 'to express' is defined as no more than to represent, to portray, or to set forth, it is clear that the use of the term in reference to art works goes further than this. In general it draws attention to emotions or feelings, and by inference can also emphasize the attitudes and intentions of the artist. In this sense, it can be suggested that if the art work is expressive it is expressing something felt or intended by its creator. As Berleant (1970: 30) points out, this has the consequence of 'making us concerned less with the object itself than with its genesis and the artist who created it.' Scruton (1974) contrasts representation, by which he means the external reality of the subject matter, with expression – the unique statement about something which the artist makes, and which can only be appreciated as a whole. In a similar manner, Wolterstorff (1980) insists that an art work expresses something which goes beyond what it may literally represent; that the reaction to this mysterious quality is immediate; and this enjoyment of the form of the work can occur even in representational works.

Langer (1957: 20) takes the view that a work of art expresses human feeling through form, and that an expressive form is 'any perceptible or imaginable whole that exhibits relationships of parts, or points, or even qualities or aspects within the whole, so that it may be taken to represent some other whole whose elements have analogous relations. The reason for using such a form as a symbol is usually that the thing it represents is not perceivable or readily imaginable.' She then uses the example of the world map to show that the same object, the Earth or the Earth's surface, can be represented equally well by two different projections. She contrasts linguistic discourse with artistic expression, and challenges the view that art is concerned with the conscious emotions of the artist. Therefore, 'the vital processes of sense and emotion that a good work of art expresses seem to the beholder to be directly contained in it, not symbolized but really presented. The congruence is so striking that symbol and meaning appear to be one reality' (Langer, 1957: 26). In this respect Langer maintains that 'Expressiveness belongs to every successful work; it is not limited to pictures, poems or other compositions that make a reference to human beings and their feelings. . . . A wholly non-representational design, a happily-proportioned building, a beautiful pot may, artistically speaking, be just as expressive as a love sonnet or a religious picture' (Langer, 1957: 59).

In art, the term 'expressionist' is usually taken to refer to a view that the artist is concerned with expressing personal feelings, not a representation of anything external. Schaper (1964) contrasts expressionist with the mimetic, that is the view of art as imitation, in which 'truth' is seen as literal or exact correspondence. She explains (Schaper, 1964: 230) that expressionism 'was a reaction to the other, earlier extreme of understanding art in mechanistic terms as a vehicle of communication, instruction and representation.' She goes on to emphasize that the importance of the concept of art advanced by Langer in *Feeling and Form* (1953) is in distinguishing between feeling and those things which are symbolic of feeling. As Schaper (1964: 231) puts it, 'Artistic configurations, though expressive, can be regarded as suggesting meanings, as indirect representations, as obliquely referring to what they express. A symbol is a kind of sign, and as such is not part of the feeling it expresses. By expressing "symbolically" it is taken to "stand for", to "call up", to "suggest" what it means.'

The breadth of this conception is important with regard to maps. Although the initial purpose in the formation of the map content is to represent, the map is not limited as a whole to direct resemblance, or specific signification. In order to convey information, the map must represent the content by an appropriate choice of both representational methods and graphic symbols. It is in the proper combination of the two that expression occurs. Through this, things can be suggested or 'called up' which lie beyond the scope of the specific meaning of the symbol. Therefore the presence of a darkened area on a map can 'express' a high value or a large quantity, even without specific map definition. In addition, the presentation of things in an organized space, although done with a different perspective to that of the picture, also makes possible the 'calling up' of associations and possible connections in the mind of the perceiver.

In this sense the map straddles the two classes of objects described by Mukařovský (1978: xxxiv), the practical and the artistic. According to him, meaning in the practical sense involves external context, whereas works of art are severed from external context. As he says, 'Their intentionality cannot be deduced from what is external to them but only from their internal organization.' To apply this to maps it is necessary to accept that two levels of symbolic signification may be present. On the one hand, the specific reference of the map symbols to aspects of the real world, which embody the map's 'information'; on the other, the composition and design of the map as a whole which may offer 'the satisfying appropriateness to one another of a thing's form and content' (Eldridge, 1985: 308). Because of this it may be possible to enjoy or appreciate the map as something more than a collection of factual references to things in space, and even the use of the map may be heightened by this feeling. Unlike the products of pure art, the achievement of this property may not be the dominant aim of the cartographer. Like the products of pure art, it cannot be predicted as the outcome of the application of any rules.

It is important to stress that this degree of aesthetic value is not necessarily present in every map. It lies beyond the point where a map can be described as 'well designed' on the basis that its use is possible by visual interpretation. Like any other artistic work, success is only achieved occasionally. It is sometimes forgotten that most great artists produced a large amount of unfinished or insignificant work, and attention is concentrated on their masterpieces. In the same way, many painters and writers have destroyed compositions which failed to express what they wished. Outstanding works of art are only a small part of total artistic output.

There are clearly many maps which offer little beyond the satisfaction of a descriptive requirement. Maps, like words, can operate at a very modest level. But in solving the problems in representation created by a given map content, the cartographer may achieve something which can yield a sense of gratification, and which is enjoyable apart from the satisfaction of function itself. Any practising cartographer must be aware of that rare moment when the final realization of a particular map brings about this sense of satisfaction. It produces what Crowther (1982: 144) describes as the ability to 'engage our whole being rather than cognition alone.' That it does occur with maps has been witnessed by many writers, including those without any primary interest in map making. For example, Watson (1982) states that 'maps are not only valuable sources of information, they can also be objects of beauty that kindle the imagination and stimulate the uninitiated and the expert alike.' Many others have stated that they 'enjoy' maps, and in so doing indicate, however unconsciously, a dimension to this experience which is extra to the functional use of the map.

Intention

As the concept of art as expressionist emphasizes the role and emotions of the artist, it also raises the question of whether art is dependent on the intention of the artist. Although it is clear that the artist, in producing any work, intends to create something which will either give personal satisfaction, or satisfy a perceiver, or both, it is hardly adequate as a criterion of what constitutes a work of art. Whatever the intention, the result may not necessarily contain any of the required properties.

The point is significant in regard to maps, for the cartographer cannot claim to be an artist in the sense of treating the desire to produce an artistic work as the primary intention in making a map. Whether a map has any aesthetic value must come from the map as a whole, not simply what it looks like at first glance. Indeed, a preoccupation with this idea may well result in a map that may give the appearance of artistic value quite at odds with its content and representation. Excessive concentration on form, regardless of content, can be a factor in producing maps that are superficially attractive but also misleading. In this sense the relationship of form and content may be quite inappropriate. This also underlines the importance of the cartographer's grasp of the subject matter and representational methods. The cartographer who sees his work as primarily accurate drawing, or who thinks of the use of colour simply as a means of making the map look attractive, may remain unaware of the important connections between content and expression. In so doing he is open to the proper criticism of superficiality, of delighting in the activity of making a map rather than the substance. It is the equivalent to the use of rhetorical eloquence to subserve trivial or misleading ends.

The criticism that cartographers make maps to please themselves, if it is founded at all, is directed primarily at this notion that the cartographer can enjoy composing something which gives the appearance of being artistic. Because it is difficult to distinguish between artistic creativity applied properly to representation, and artistic creativity misapplied to superficial attraction, the subjective or individual element in cartographic work is often regarded with suspicion or even hostility. It can be used as an excuse to deny the true role of creative expression in the production of a map. Yet it is difficult to imagine anyone being drawn to cartography, having a desire to become a cartographer, who did not feel that cartography was truly creative in some way.

Immediacy

Although this property of aesthetic value can be referred to, it is difficult to describe in concrete terms. In relation to visual art, linguistic description tends to fall back on metaphor, for there is no simple linguistic equivalent to the sense of appreciation aroused by a work of art. Crowther (1982: 144) attempts to define it as follows:

> To use a sign-system analytically – as in the traditional intellectual disciplines, or indicatively – as in the propositions of ordinary language, involves a bare and essentially cognitive operation with signs. There are occasions, though, when we use signs in a much richer way; in a way that seeks to construct or re-construct some aspect of the world, in all its sensuous immediacy, i.e. as it might be encountered in perception itself.

The expressiveness of any art work is something responded to by the viewer. Langer (1957) maintains that this cannot be pointed out; it is apprehended, and in this sense is intuitive. If it takes place at all – and in this respect viewers are not uniform in response – then it happens spontaneously and immediately. Langer also suggests that this intuitive response is not a special gift, limited to a few, but is an unrecognized part of everyday experience: unrecognized because it does not usually lead to any verbal description. As she says (Langer, 1957: 68) 'artistic perception, therefore, always starts with an intuition of total import, and increases by contemplation as the expressive articulation of the form becomes apparent . . . the import of an art symbol cannot be paraphrased in discourse.' The same point is made by Pye (1978: 108): 'When we see a true work of art the experience of beauty not only illuminates but in some way as it were irradiates and transmutes whatever our intellect apprehends from it so that we become sharply conscious of much that is quite beyond the reach of direct verbal expression.'

This reward of contemplation is not an equivalent of the serial searching for information on a map referred to earlier. The import is apprehended at the beginning; it grows in satisfaction as it is unconsciously explored. The first encounter with a great work of art previously unknown demonstrates this: the sense of delight, the feeling that something intensely satisfying has been discovered, the recognition that this object will continue to provide pleasure in contemplation.

With maps, this sensuous immediacy might seem to be relevant only to those aspects of the 'real' world that are visually apparent, such as the representation of relief by colour and shading, or the symbolic use of blue for water. But there are many other kinds of map, including those dealing with invisible or intangible phenomena, which also have this property. The presence of a scale of graded colours in a map of temperature not only reinforces the visual distinction between different categories or values, but presents this in a manner which offers to the senses a metaphorical equivalent. In so doing, it is more immediate, more concretely apparent, than an equivalent set of lines, which have to be interpreted in a wholly cognitive manner. The more immediate the sensation, the more it is tied to a distinct interpretation of the subject matter by the map maker, and the more likely it is to arouse feelings about its appropriateness on the part of the map user. This sensuousness is always apparent in the perception of colour, as is seen in the response of map users to what is 'appropriate' and what is not. In this respect, the map is rhetorical not because the use of colour is misleading, or persuades the map user to believe something which is untrue, but because the effect it achieves reinforces the information being represented.

This appeal to the senses can operate in many different ways. An engraved topographic sheet of the early nineteenth century, with its fine lines and its bare simplicity in black, is visually very different from a multicoloured map of any kind. Yet it too can provide a sense of enjoyment, and one separated from any function, for it is no longer a representation of the world as is, but only of the world as it was. True, it can be used as a source of information about its period; but it can also be enjoyed for its own sake. It is probable that a cartographer would find enjoyment in the craftsmanship more than a non-cartographer, just as those people who delight in paintings may have a deeper level of satisfaction than those who do not. But the satisfaction of perceiving such a map, created within the limitations of a given medium, is akin to that offered by a pencil sketch or a line and wash drawing. This satisfaction is based on an appreciation of the skill with which a representation can be made with great economy, yet still offer something to the imagination.

Creativity and imagination

It is commonly agreed that any work of art is unique and individual, and that therefore it is an original creation. This stands even though specific works of art may be grouped into certain styles or historical epochs, as sharing some common properties or conventions. The nature of creative originality has been discussed under many headings. It has been considered in the sense of trying to establish what makes a work of art original, and also how creative activity takes place.

The question is more involved than may first appear. In many types of human activity, especially those in which design, marketing and the production of consumables are important, the desire to produce something different, something new, is a recognized part of the activity. So much so that many formulae have been described for assisting this, most of them based on the assumption that if it is a necessary activity, it should be possible to analyse it as a process. Even so, as Hospers (1985) points out, it is not simply uniqueness, but memorable uniqueness that distinguishes a work of art. Merely making something different does not necessarily achieve any aesthetic property, as the study of fashion makes clear.

At the same time it is clear that creativity is also vitally important in science, for the outstanding scientists are those who discover new truths or ways of seeing things. It is easy to confuse problem solving with creativity. In design work, problems usually arise within the overall framework or concept; they are centred on the fitting and balancing of the individual parts. True, realizing where the problem lies is important, and finding a solution may involve going beyond the normal conventions. But this does not correspond exactly to the origination of an art work before the point is reached where there are specific problems to solve. A painter may struggle with some aspect of a picture, the way in which some element should be rendered, and become exasperated by lack of success. But this can be distinguished from the origination of the concept of the picture itself.

Some writers, such as Buttimer (1983), have pointed out the close correspondence between originality in art and science, to the point of attempting to analyse the characteristics of people who are leading members of the two groups. They tend to emphasize the importance of divergent thinking – the capacity to go beyond the conventional and expected; and the high degree of confidence and conviction common to both scientists and artists; the belief that they can do what is needed. They also tend to emphasize the aesthetic satisfaction of advances in science; that mathematicians speak of 'elegant' solutions, and advanced physical theories have a 'beautiful' simplicity. But how ever it is reached, the common factor is that both are occasionally capable of

217

creating something not previously conceived, and that this cannot be revealed by any systematic analysis, because the factors themselves remain unknown. Original works in both science and art seem to be inspired by something that defies normal explanation. At the same time it is necessary to distinguish between great discoveries and artistic masterpieces, and the more general levels of art and science. Although most maps, like most scientific experiments, do not lead to outstanding achievements, they can include a dimension which lies beyond ordinary contemplation or the following of production routines.

Buttimer (1983) points out that much of the research in this field, especially by psychologists, has concentrated on cognitive processes in problem solving, and defining creativity in terms of products. On the other hand, sociologists are more concerned with external social factors such as the effect of cultures and institutions. Most of this tends to reflect 'the central biases of the West, viz., judging creativity in terms of visible product, emphasising the intellectual over other human faculties, the knowledge of active problem solving rather than understanding or contemplation of mystery' (Buttimer, 1983: 20).

Despite the many attempts to describe the stages through which any original work develops, what lies behind the initial concept or idea remains mysterious. Buttimer (1983) refers to five phases that may occur in the creation of a new work, whether this is artistic or scientific: these are initial stirring, intersubjective discussion with others, analysis and research, synthesis and articulation, and communication. Many people have pointed out that the initial idea seems to arise spontaneously, and is rarely produced by deliberate action. Often an idea has to be left or put aside before it takes on a more concrete or developed form. As Buttimer (1983: 131) puts it, this is a 'period of concentration and then a deliberate abandoning of the quest for a while.' This suggests that something is going on in the subconscious, which lies outwith any conscious effort.

Hausman (1981: 77) pursues the problem of creativity at a deeper level, maintaining that 'creativity occurs on condition that a new and valuable intelligibility comes into being.' He continues, 'since its intelligibility is unprecedented, it is underived and is something that was not predicted. But also, its intelligibility is discernible in a structure that *could not have been* predicted' (Hausman, 1986: 82). Thus Hausman distinguishes this from examples of art which are based on a convention and thus derived, and things which are simply new or different but do not have intrinsic value. Being novel or eccentric is not sufficient in itself. There are many maps, such as the United States topographic map series, perfectly satisfactory in their way, which have been based on accepted conventions; and many others have attempted to be novel or exciting without necessarily being successful.

The subconscious phase is one that appears to have links with problem solving. Once a problem is detected, the eventual solution may well lie in approaching it from another direction, or taking a different starting point. The ability to do this is often dependent on 'sleeping on it', or in some way giving time for something to develop. It is evident in many cartographic problems, in which the initial phase of looking at the material and the possible solutions does not produce the desired result. In this case it is the ability to think up some different approach that will decide the outcome. There is a link here in the sense of originality, but it still does not remove the difference between problem solving and the creation of a new concept or means of expression before the content has been articulated.

The design and production of a map can rarely be regarded as equivalent to the conception and development of an artistic work in fine art, because the requirement and the conditions are affected so much by non-cartographic issues. Cartographers do not sit around thinking up ideas for new maps: they produce what the organization or the client requires. The fact that map production takes place within limitations is not in itself the chief difficulty. Many works of art have been produced for clients on agreed subjects or topics: the artist was able to use these as the basis for creative work. Many great artists have acted under the pressure of fulfilling commissions; for example, Mozart would compose a piece virtually to order if offered a suitable fee.

The chief restriction in cartography is that so often little attention is given to the composition and design stages. Most map designs follow conventions or accepted practice, with slight variations, because production under pressure and the need to satisfy the client inevitably leads to this. Few clients are willing to support anything so intangible as 'creative design', and in most cases the design specification is regarded as the routine task, not as an opportunity to produce something original. Most new map designs, how ever much they are aggrandized in advertising, are no more than slight stylistic changes on familiar material. The objective of satisfying the customer, directly or indirectly through a client, usually confines the product to within the realm of the expected and the usual.

Very little art would be produced if artists were content to give customers what they wanted, or thought they wanted. Indeed, the world of art is littered with the recollections of great works that were not appreciated in their time, because the artist did not work within the confines of what was then accepted. There are many examples of a new work being greeted with derision on the part of the public, such as Berlioz's *Symphonie Fantastique* and Whistler's *Nocturne in Black and Gold*. It is the artist's insistence on doing something different which leads to major developments in the arts, and which in turn establishes new conventions.

Imagination

The term 'image' and its derivatives are widespread in cartography. Single elements in map construction are described as images. They are concrete manifestations of the information in a given form. Visual imagery – the notion that the map user may construct a mental image or picture – is regarded as important as one aspect of map interpretation. Although such visual imagination, stimulated by the map, may clearly operate where the map represents observable reality, at a scale which makes correspondence between map and external world possible, the construction of images for phenomena which are never directly perceived must operate in a different way. In this respect imagination is of consequence both for the map maker and the map user.

It is clear that the creation of a new idea requires imagination, how ever uncontrived, for it has to be brought into existence without reference to what is already known. Yet the power of any work of art is that it expresses something which can also appeal to the imagination of the perceiver. The expression itself may rely on the use of metaphor; and describing the nature of an art work's appeal is heavily reliant on metaphor, for language has no direct means of embodying such individual sensations. Thus most of the terms used to describe the aesthetic properties of a picture or a musical composition are essentially anthropomorphic; they describe, not what it is, but how it appears to affect the perceiver. So a picture may be serene, aggressive, lively, gay, detached; a musical composition may be mournful or exciting. This takes place because of the way in which the perceiver's imagination is aroused.

In cartography, the use of colour is often metaphorical. According to MacCormac (1971–72: 240), 'Metaphor is a linguistic device in which words are juxtaposed in such a manner that a literal reading would produce absurdity.' If this is transposed to a map, then for example the application of the colour purple to an area described as having a high annual precipitation is representationally 'absurd'. It seems to be an appropriate description because the strength and visual impact of the colour is associated with the idea of a large amount.

This opening up of possibilities beyond the literal facts of representation is certainly one of the properties of maps. Those early and inaccurate representations of 'new worlds' and recently discovered continents, with attendant monsters, certainly excited the imagination of the map viewers, and indeed were intended to. Understanding that the flat map of the world represents the surface of the Earth requires imagination as well as a grasp of projective conventions. Even the contemplation of a topographic map of some area previously unfamiliar can lead to the imagination of its form and character, and this can go beyond the bare concept of its physical appearance. This acts partly by the intrusion of what is already known as a basis for extending experience, and partly on the free play of the imagination, which appears to go on without control. The higher order, or second level, interpretation often referred to in descriptions of types of map use is also dependent on the ability to imagine not only surfaces or features, but possible connections between them.

Although imagination is one of the most important distinguishing features of humanity, its development as a visual sense is also affected by use. in this respect, as Sparshott (1981: 52) points out, 'The artist, the poet, is to be known by his sustained habit of attention.' What the artist attends to, at least with the representational arts, is what things look like, and the effect this has on the imagination. Price (1981: 100) makes the dry comment that 'We live in the most *word-ridden* civilisation in all history,' and that 'our whole educational system is directed to the encouragement of verbal thinking, and the discouragement of *image* thinking.'

The ability to use imagination is important to the cartographer, both in the sense of generating original ideas, and also in being able to imagine the outcome of a decision in terms of graphic appearance. This often requires sustained attention, and this attention to the ways things look is an acquired habit, even though it may depend on talent. The most important thing any beginner in cartographer has to develop is the ability to use visual imagination as a testing bed for ideas of representation, and to develop this also by spending time attending to the visual properties of maps. The undue concentration on methods of representation as verbal concepts (contouring, shading, proportional symbols) divorced from the vital connection with their actual expression in a map, indicates the difficulty of achieving a full development of 'image thinking'.

The need to give attention to the creation of the object also helps to explain why it is that artists are more concerned with doing than with explaining how they do. As Sparshott (1981: 52) puts it, 'What poets most evidently have in common is not a mysterious contact with secret springs, and certainly not any shared mental process, but simply a steady application to the writing of poetry.' It is this constant application which enables the artist to understand the effect of doing certain things, and it develops to the point where it operates subconsciously. It is quite evident, for example, that John Donne was technically expert in the use of accent and change of rhythm. It is highly unlikely that he consciously thought of these devices in the process of writing.

Most books on painting and drawing emphasize that the beginner or student must practise all the time; must develop the habits and experience of the artist by deliberately looking, representing and expressing. This embedded experience is also what the practising cartographer depends on, and it is essentially different to what can be acquired by theoretical study. Properly applied, it lays a foundation of knowledge of how effects can be obtained, which provides the means by which the creative imagination can work.

Aesthetic properties and maps

There are two major aspects of aesthetics in so far as they apply to maps. On the one hand is the fact that as the map is a graphic creation it inevitably affects the map user by its appearance, and this cannot be avoided. On the other hand, it is possible to regard maps as of interest in themselves, apart from their ostensive functions.

The first factor is critical in discussing the role of art in cartography. The notion of a 'good' map certainly involves two sets of qualities: its informational content, selection and treatment; and its aesthetic appeal as part of this treatment. This 'satisfying appropriateness requires the unity of content and expression. Like other forms of art, including fine art, this unity cannot be achieved by regarding design as a separate function, or treating the visual quality of a map as something that can be considered separately. As with all forms of rhetoric, it can be abused or misapplied.

One of the major difficulties in studying cartography, as with many other disciplines, is that the processes involved in the preparation and construction of a map have to be dealt with as a series of separate bits, usually linked in a time sequence. So the compilation is made first, the methods of representation chosen, and then the symbol specification is 'designed', and the production processes planned and executed. But if these stages are treated as separate operations it is unlikely that the desired unity of form and content will be achieved. The generalization of source material is itself affected by the intended or presumed design. The selection of place names is affected by the selection of content, but on small-scale maps the need to name places with particular lettering sizes will itself affect the content selection. Therefore content and design have to be thought about together. A good cartographer will have, at least in imagination, some idea of what the final appearance will be, and in this sense manipulates both the content and the expression towards this objective. Given that much of this lies only in the imagination until the map has been produced, it is not surprising that few maps are satisfactory on the basis of the first attempt.

It also underlines the importance of the conception of the map as a combination of science and art. As such it is necessary to understand that the the map is a mixed or hybrid form, and this underlies much of its interest and fascination. It is true of all objects designed for use, and as Pye (1978: 95) observes, 'If design is a problem-solving activity, it is also an art. Let us forget that all too common phrase, "Art and Design", suggesting as it inevitably does that design is distinct from art.'

The idea that maps are of interest in themselves is perhaps more unusual, but there are many people who have voiced the opinion that they 'like maps'; that is, they simply find them interesting apart from any specific use. It is most evident perhaps in those cases where maps are treated as antiques, as valuable objects produced in the past. Such maps are often displayed or exhibited, and therefore presented as interesting examples of human skill and artistic expression, as well as historical curiosities. It is also clear that this interest is very much in the eye of the beholder, and is limited to a few people, just as the enjoyment of paintings is also limited to a small proportion of any population. As Pye (1978: 122) says, 'each man's response to art very obviously differs, according to his temperament, from that of the next. It is idiosyncratic.'

To examine the question of why maps are interesting in this general sense, it is necessary to go beyond the bare idea of aesthetic contemplation. As maps deal with such a range of natural and social phenomena, they extend the realm of human experience. This is probably close to the modern interest – as revealed by many television programmes – in documentaries, or in films on nature and natural phenomena. It no doubt relates to our endless curiosity about the world we inhabit. Maps give form to knowledge, and stimulate the imagination. The first exploration of the colours and patterns of a geological sheet can be as interesting as the first reading of a good historical biography. There may be no overt reason for so doing, and the intention is not essentially to acquire information: the act is pleasurable in itself. The enjoyment is a property of the form of expression as well, just as a good biography is also a function of the quality of the writing. It is unlikely that the observer will be conscious of this aesthetic element, any more than the reader will actually concentrate on the linguistic expression of a text. It will be subsumed within the overall satisfaction.

That any map can be enjoyed solely on the grounds of its aesthetic appeal is more debatable. It is possible that the enthusiastic cartographer may revel in the imagination, understanding and technical skill shown in a particular map, without being conscious of either the subject or the area. In most cases, it is more likely that the cartographic enjoyment will lie in appreciating the skill with which the subject matter has been expressed, that is, with the way in which information, composition, representation and design are united in their appropriateness. To look at a map in a purely contemplative sense would reduce it to the level of an abstract work of art. Bearing in mind that the map is always about something, that it is a representation of something, this would seem to be an unlikely feat.

CHAPTER 15

Cartography as a graphic art

Although many texts on cartography make some general reference to art, or the need for the cartographer to have some kind of artistic ability, the possible connections between cartography and art, or between the practices of cartographers and artists, are rarely examined in specific terms. Whether such ability is central to cartographic activity, or simply a means of achieving some superficial improvement in a map, is not even considered.

It has been argued already that as a picture and a map are fundamentally different, any parallels between cartography and art must lie elsewhere. There are a number of possibilities, including the characteristics of the mediums within which they work, the modes of expression, the nature of study and practice, and the responses of the perceivers and interpreters.

In painting, the physical nature of the work – whether it is in oils, water colour, crayon or a hybrid such as line and wash – helps to define a major characteristic and also influences choice of subject and method of working. Although such distinctions might be applied to manuscript maps, in general virtually all published maps are intended for reproduction by printing, and therefore the question of medium does not directly arise. Although many representational drawings are intended for publication and therefore reproduction in print, normally these are produced independently of the reproduction process. In this sense, the procedure of colour-separated production typical of traditional analogue cartography, with its combination of manual and reprographic processes in generating image components, or the generation of a digital map on a screen as a sequence of operations, has no parallel in other graphic arts, apart perhaps from the autolithography of the nineteenth century.

Cartography and the modes of expression

Because of the preoccupation with individual symbol design, and thus the separation of design from production, the effect of using the different modes of expression is never discussed as such cartographically, despite the many references to line maps, colour maps and black-and-white maps. Yet these modes of expression are regarded as fundamental in drawing and painting. This in itself is interesting, because of course the same means are available to both cartographer and artist.

In 1743, Osborn published a book on painting which sought to describe both the principles and practice of painting as a fine art. In the chapter on design he distinguishes between design in the sense of the idea or entire thought of the work (composition or planning); design as representation, the study of forms in real life; and design as 'outlining', or drawing (Osborn, 1743: 78). The last of these three treats design as one of the three essential components of the art of painting. Subsequently, he distinguishes design from both light and shade and colouring: 'the incidence of lights, the artifice of the *claro-obscuro*, local colours, their agreement and union with one another, their aereal perspective, and the effect of the whole together' (Osborn, 1743: 269).

In many other treatises the distinction is made between delineation or outlining, chiaro-obscuro (light and shade) and colouring, but they all start with the emphasis on outlining, or drawing. Osborn (1743: 79) maintained that 'design must be learned before everything, as it is the key of the fine arts, as it gives admittance to the other parts of painting.' Hamerton (1882) explains that 'All drawing began, in primitive times, with simple outline; and the next stage was to fill in the spaces so mapped out with flat colour, but the outline was still preserved for a long time in all its hardness of definition. The first notion of drawing which occurs to man is to mark out the shapes of things in profile with a hard line.' In several instances the qualification is made that drawing may be a preliminary to painting rather than an end in itself, or that the outline drawing is the preliminary to the complete drawing, with its tonal shading and three-dimensional perspective. As Osborn (1743: 78) says, this circumscribing of exterior forms is 'a sort of creation, which begins to fetch, as out of nothing, the visible products of nature.'

The emphasis on drawing in art education is discussed by Meeson (1972: 277). He points out that 'Drawing is thought of, characteristically, as the one essential skill which distinguishes the artist from other men, and indeed it holds such a fundamental place among the skills of the artist that until comparatively recently art education usually meant, quite simply, lessons in drawing.' He goes on to comment that after the Renaissance, drawing was connected to the idea of understanding the objective world, and could be used in the service of science; but that this was followed by a different conception of drawing – that of a 'pleasing manner of performance leading towards virtuosity.'

The notion of a map as a 'drawing' has deep roots in both cartography and the opinions of map users. The idea of a cartographer, for most people, is a person who draws. This fundamental requirement for drawing ability has always characterized the cartographic draftsman, so much so that as this requires developed skill, it has often been regarded as the 'artistic' element in cartography. Although the work of the draftsman is associated more with drawing to rule, following measurements, or simply precise copying, a high level of drawing skill was essential during the long period when map production depended on manual performance.

Delineation

Delineation, the drawing of objects in outline, is regarded as the basic initial process in graphic representation, and it is evident in all maps. It is referred to frequently in accounts of the history of cartography, both in relation to representation and technical methods. Wood (1985: 82) points out that 'The success and continued use of engraving as the primary technique apparently stems from the ease with which linear handiwork could be done by the draftsman.' Skelton (1962) refers to an early map by Yeakell and Gardner as 'the earliest printed delineation, in English cartography, of an extensive tract of country.' In many cases, the entire map is a delineation, using only a combination of lines and outlines for linear features and areas, and outlined shapes or solids for point or area symbols. This operates so extensively that 'line' maps are often referred to as a basic or generic class. This expression also has technical connotations. The line is the simplest graphic image to reproduce in print, and even after lithography took the place of engraving as the principal means of map printing, the ease of line reproduction still had a strong influence on map style.

As is well understood in art, outlines do not actually exist in nature, and the notion that any real three-dimensional feature can be represented by a line depends on convention, a way of seeing things or a way of conceiving things (Fig. 40). As Pope (1949: 58) points out,

> All delineation is of course abstract, and it has little to do with the popular notion of imitation of superficial effect; but because line drawings are easily read by almost anyone, this abstraction is usually not thought about much. There are obviously no lines in nature; there are only areas of tone distinguished from each other by contrasts at their edges. The extension and shape of objects are ordinarily most strikingly marked by these edges, and so when we indicate the position and shape of these by lines we express the general form of objects to some extent.

Cartographically the convention has a basis in the nature of the topographic map itself, and the way it is comprehended as a representation. Linear features such as rivers and roads are essentially encountered as features or interruptions at ground level, represented as lying on the two-dimensional map plane. It is not that they are not three-dimensional but they can be thought of as features in a two-dimensional surface. Therefore the representation of river systems, roads, hedges, fences and walls (and by extension all types of boundaries) does not depart radically from the way in which the map is understood at a simple level of interpretation. They correspond quite easily to the idea that the map shows 'where things are on a

Figure 40. Delineation: (a) the outline sketch; (b) the survey plan

two-dimensional surface. The same principle applies to specialized maps where the subject matter – such as population or trade – can be thought of as existing at ground level in a general sense. Where they run into difficulties, of course, is when a significant class of objects covers only a small extent at ground level, and cannot be shown in outline at scale. Exaggeration on the basis of significance has always been accepted as a necessary device in cartography, as it has in painting.

Unlike a representational drawing, lines in cartographic representation are specific, no matter how abstract they are in concept, because they are themselves denotational in intent. All map symbols have to be given form in the first place, either as lines or outlines, or as point symbols. The line itself has to be determined in terms of its three properties of form, dimension and colour. In painting, the outline is seen essentially as a means of revealing shapes and surfaces; in cartography, the use of the line is much more complex. The difficulty which this requirement introduces is that initially at least line symbols are conceived as separate representational devices, dominated by the relative importance and nature of the feature being represented. Consequently the relative effect of one line in relation to another may be ignored or given little attention. In a drawing, where the use of line is more flexible, and not mentally limited by any requirement for

227

representational consistency, lines are varied as needed, and their combined effect is perceived as the work is gradually produced. Lack of awareness of the combined effect of lines is often shown in simple illustrative maps. Whereas an artist is always conscious of the effects of the graphic forms being used, some maps make little use of the possible variations in line gauge and continuity. The act of placing the devised symbol in the right place is taken as conclusive, regardless of its contribution to the composition as a whole. The line image has to carry too much, and the result is confusion.

The other factor in delineation is of course the nature of topographic survey itself. The surveyor operates by measurement. For this, fixed, tangible positions are needed, and therefore naturally there is a concentration on those elements which can be clearly defined at ground level. The 'line' of a hedge or fence, the edge of a road, the bank of the river – these are the basic elements that are measured. In this sense the forms and surfaces are extracted by delineation, and it is this delineation that provides the survey plot. Therefore the initial product of a topographic survey, whether produced by ground observation or through the interpretation of aerial photographs, is inevitably a series of delineated points, lines and outlines.

The same procedure operates with the preparation of derived maps and many specialized maps. The initial compilation is basically composed of marked points, lines and outlines – the essential informational content in rudimentary form. Even those areas which in the final version will not have outlines have to have their boundaries described by outlines at the production stage. This line image may then be developed and elaborated in design, and in its completed form, with names and other elements added, serves as the basis for the production of the primary images.

In several other kinds of map, representation can be limited to the range available in line drawing. Many statistical maps using proportional symbols contain only lines and outlines, as did nautical charts for a long period. But when the only drawing device available is the line, the medium places a strong obligation on the cartographer to define both character and importance by carefully controlled line forms. The line image can be heavy or delicate, detailed or simple, with strong contrast or little contrast.

Light and shade

The second element in painting, the use of *chiaroscuro* or light and shade, is more difficult to define. The term was introduced when Italian painters began to 'model' three-dimensional surfaces by highlighting and shading, but initially this was limited to individual surfaces. With the development of pictorial perspective and the representation of scenes as perceived in nature. then the use of a specific light source was adopted, which affected both local tone and three-dimensional modelling. A painter like Rembrandt could choose a light source for the purpose of dramatizing a scene, making full use of strong contrast between light and dark areas. It was consistent to the picture, but did not necessarily match any natural view. This is different to the introduction of a specific light source in a landscape or view used to suggest three-dimensional form and depth, or the way in which objects would actually appear. In such a representation the shadows cast by objects in relation to a light source are an important part of the total information, reinforcing the impression of a perspective view. Local tone, which suggests the position of the surfaces in relation to the light, is distinct from shadow.

The addition of shading is a natural progression from the initial outline drawing. Howard (1848) first explains that

> a mere outline, or contour, if well studied, may give a characteristic idea of the figure of any object, by shewing its general proportions and lateral boundaries, but to convey its anterior appearance, its various projections, planes and hollows, the contours must be filled up with a scrupulous copy of every gradation of tone observable on its surface, from the highest light to the deepest shade.

He continues that 'in super-adding to these, light and dark which do not necessarily belong to the subject . . . and deciding on the intensity, quantity and arrangement of these tones, chiaroscuro becomes essentially inventive or ideal, and one of the most powerful agents of the painter.' As soon as areas or surfaces are distinguished, by whatever method, then some degree of weight or emphasis is automatically given (Fig. 41).

Figure 41. Line shading in drawing and line patterns in cartography

Although applied to different ends, it is quite normal in cartographic design to devise the line symbols first, and then to consider the addition of patterns or tints to the 'outline'. This is particularly true in black-and-white maps, where the addition of patterns or tints of the black may be critical in making the map legible.

Shading

Another difficulty is the vague connotation in the graphic arts and cartography of the term 'shading'. In many accounts, the addition of any type of image to a surface is described as shading. In this sense shading can be regarded as the addition of a line or point pattern, or a continuous grey tone, or even a continuous colour. For example, Wood (1985: 82) in describing the development of map engraving, says that 'the fine point of the engraver's burin allowed him to dot or flick the surface to produce the delicate effects of shading where desired.' Such a characterization of a surface, so often necessary in a map, is primarily for purposes of description and identification; it does not necessarily relate to any visible aspect of a feature. In this respect, the addition of any area symbol, whether a repeated series of small point symbols, a pattern of lines or other forms, or a continuous tint, may be described as shading, even though the initial function of such symbolization is to describe the area according to some classification, not to indicate its visual appearance in respect to a light source. It is also unfortunate, because in the description and analysis of colour, the term 'shade' has the specific meaning of referring to a devaluation of a hue by the addition of black or grey, and is therefore the opposite of 'tint', which involves the lightening of any hue by making it more transparent or by adding white.

It is also the case that in many maps, linear features and point symbols are present in areas which are also shaded. The treatment of the two in combination is of fundamental importance. This is where the visual appraisal of the map as a whole is so critical. It is hard to imagine any painter adding elements to a landscape painting by a series of successive additions, one on top of another, without considering the effect of one on the other. Yet in practice this can happen all too easily in cartography. Because of its construction methods, a map, like a painting, has to be assembled piece by piece. But whereas a painter works to a conceived end, to which all elements in the painting contribute, maps may be put together as a collection of separate bits, defined by individually chosen symbol specifications.

Shading and the use of white

In a map produced purely by delineation, most commonly using black only, an essential difficulty is that maximum contrast is used at all points, i.e. the contrast between black and white. Although some visual separation can be achieved by variation in line thickness and continuity, the constant high contrast provides an image which tends to lie on a single plane, and therefore is graphically limited. In drawing, the initial outline is normally regarded as no more than the first stage in the composition. Once established, it is then worked on by developing contrast.

The addition of area symbolization, of any type, to a line image immediately changes this, not only because it clarifies classified areas, but because it makes possible the use of white as an element in the composition. Instead of being simply a neutral and unoccupied background, white areas can be contrasted with those which are patterned or shaded.

Unfortunately, because the construction of a map tends to proceed by the addition of sets of symbols, the importance of the use of the white background tends to be ignored. Because the cartographer may think of the image primarily as a space in which things are placed, the effect of the unoccupied space is missed. Yet for many painters, almost the first thing they do is get rid of the white background. The contrast of light against dark is just as important, and just as useful, as the contrast between dark and light.

Colour

The third component for the painter is colour, and it is not surprising that it is the element which occupies so much attention in works on artistic practice and theory. By far the most complex of the graphic variables, it is employed for both its representational and its emotive aspects in equal measure. Howard (1848) states that 'Colour, like chiaroscuro, may be treated either as a property to be found in the local hues which naturally belong to every object (that is, as merely imitative) or as it forms part of that ideal whole which he has conceived in his own mind, and seeks to call into existence on his canvas, and is altogether inventive or theoretical.' The affective power of colour is revealed in the terms artists use in its description. Bright, sombre, warm, cold, transparent, advancing, receding, neutral, bold – the whole terminology is a mass of metaphors. Although in cartography colour is used primarily for contrast, and to some extent for its naturalistic associations, this metaphorical application of colour is also highly important.

Even in the earliest coloured manuscript maps the use of colour was based on a variety of premises. Its graphic value as a means of distinguishing one thing from another – primarily contrast in hue – was rapidly adopted in conventional symbols. In portolan charts, lines of different compass bearings were distinguished by different colours; important place names were shown in red, others in black; rivers were shown in blue. Even at this stage, the power of colour to be both descriptive and affective was deliberately employed.

Dainville (1964: 334) shows that in early manuscript maps and hand-coloured engraved sheets, the conventional use of colour had a long history, largely developed from painting and medieval cosmology. He suggests that 'Red above all is the colour constantly employed since the Middle Ages to indicate towns and buildings on small-scale maps.' The point is supported by Andrews (1975: 254), in describing the early large-scale plans in Ireland: 'By a longstanding tradition buildings were either carmine or grey, the former indicating slate or stone roofs, the latter thatch. Water was blue, roads and foreshores buff, while the interior and exterior walls of public buildings were picked out in deeper red.'

The early printed maps depending on either an engraving or a woodcut for their reproduction could not provide a satisfactory means of introducing colour, and so for a long and formative period cartographic design was essentially achromatic. The possibility of making printed maps in colour was one of the great advantages of lithographic printing; it brought about a revolution in cartographic representation of greater significance than any subsequent event. The lithographic reproduction of paintings – in which the coloured original was separated by eye by the lithographlc artist – was never very efficient. As maps were conceived essentially as delineations, the practice of colour-separated production led naturally to the establishment of a style in which line images were either printed in colour, or colour was added to outlined areas. So strong is this tradition, that it is still unusual to find a multicolour map which has been conceived as a composition in colour in the full sense.

The major exceptions are those specialized maps in which the whole of the land area is characterized by areal symbols, such as geological or land use maps. Given that many of these may contain 60 or more different symbol classifications, it is clear that a systematic use of the colour variables is essential if the objectives of contrast and balance are to be achieved.

The growth of scientific understanding of both light and optics from the eighteenth century onwards, the interest in landscape painting and the naturalistic use of colour, ensured that colour was a topic of general, as well as artistic, interest. Even a philosopher like Alison (1811: 295) could write that colours

are expressive of many very pleasing and affecting qualities. These associations may perhaps be included in the following enumeration: 1st, such as arise from the nature of the objects thus permanently coloured. 2ndly, such as arise from some analogy between certain colours and certain dispositions of mind; and 3rdly, such as arise from accidental connections, whether national or particular.

He goes on to describe particular colours as cheerful, melancholy, pleasing, bold, mild – all being 'terms obviously metaphorical'. In the eighteenth century, as Dainville (1964: 336) explains, there was a preoccupation with aesthetics and good taste, which also affected what was regarded as suitable for the colouring of a map. As he puts it, 'Geography and topography participated in the artistic tendencies of the period. One can detect in the hand drawn plans, their design and colouring, the terms, familiar to contemporary amateurs, of drawing, painting and shading.'

Line and tint

The addition of colour to outlined areas by adding either patterns or flat colour is one of the oldest devices in cartography. It has also been used very widely in art types divorced from the school of modern Western perspective. For example, Japanese popular painting had a strong influence on the style of Toulouse-Lautrec, especially in his lithographic posters. Pope (1949: 77) describes this as the mode of line and local tone, and shows that it was dominantly used in Chinese and other Asiatic painting. Although the form was expressed by line, 'the local tone of each object is also given by means of pigment spread over its field in painting.' And he continues, 'moreover, arbitrary gradation may be used in order to accent edges and thus assist the expression of form by lines.' As he explains, 'This has nothing to do with light and shadow; it is merely a way of distinguishing planes forward and back, and is perfectly abstract.' This use of local tone is much closer to the way in which area colour is generally applied in cartography. A more subtle application is in edge reinforcement. The use of the vignetted band on many charts is a modern example of the same technique.

The combination of line and colour wash or tint was frequently described in nineteenth-century texts on painting as a significant and valuable method. In the period when the engraved map, printed wholly in black, was dominant, colour could only be added by hand, and adding water-colour washes to delimited areas was a recognized craft in the nineteenth century.

Hamerton (1882) introduces his chapter on this as follows: 'The essential difference between auxiliary washes and independent work in water colour is that such washes are only used to help linear work by giving it a sustained surface of shade; they are not of any great complexity in themselves,' and he continues, 'The line carries the wash and gives shape to it; the wash bears out and fills up the line work, and gives it consistency by bringing its scattered elements together.'

The addition of an image to any surface alters the arrangement of lightness and darkness in the overall representation, regardless of its type. Graphically, it gives continuity to defined surfaces, it assists in their identification, and its treatment is a consequence of the degree of significance given to the feature in the composition as a whole. By giving visual continuity to areas it helps to overcome the problem of detecting and correctly identifying them in map interpretation; it makes possible the visual separation of distinct features which overlap; and it provides contrast against the presence of linear features themselves. As with all graphic methods, lack of control is equally likely to introduce conflict between different features, rather than their visual separation.

Colour and symbols

Despite the fact that modern texts on cartography usually devote a considerable amount of space to the analysis of colour, colour theory and the application of colour in map symbols, there is surprisingly litt' literature on the subject of colour in cartography as a whole. This is in considerable contrast to the history of painting. Since the Renaissance in particular, artists, philosophers and poets have been interested in exploring its intriguing mixture of cognitive and expressive qualities. Throughout the centuries there have been discussions of the relationship between colour and both objective and subjective impressions, ranging from the association of hues with emotions, moods and religious representation to heraldic symbolism. The difficulty is that although it is possible to describe in words the difference between the informative, expressive and affective uses of colour, these are not distinguished in perception. In this respect, the visual impression of colour is just as 'real' as any denotative meaning.

The emotive aspects of colour were considered at length in the period of Italian Renaissance painting. Red was seen as aggressive, agitating, disturbing; green as soothing; yellow as warm; blue as cold and negative. The different hues were linked to music as a means of arousing emotions. In this context, white, black and greys, the achromatic colours, are regarded as cold and gloomy. It is evident that although colour can be used on topographic maps in a rather generalized naturalistic sense, as well as to emphasize important elements, the metaphorical use of colour is particularly important in specialized maps, especially those which do not represent any visible or tangible reality. It is the relationship between colour and conceptions of value and degree which underlies most of the practice of colour use in such maps. Generally speaking, fully saturated and darker is equivalent to either more of, or more important than; and less saturated and lighter is equivalent to either less of, or less important than. In a map of precipitation, small amounts are represented by pale tints, frequently in blue or green; large amounts by darker tints or solid colours, often moving from blue towards a darker purple. On the other hand, in maps of temperature variations, the relatively cold areas are coloured blue or purple, and the warmer areas yellow, orange and red. These also use naturalistic associations for effect, and therefore seem to be an appropriate basis for symbolization. Such scales of colour gradations are intended to be perceived as representing difference in quantity, which indeed they do. They are fundamentally metaphorical.

More important features, especially if they are small, are usually emphasized by high contrast hues (red, orange and violet), whereas those lying at a secondary level, especially if they are large in extent, are represented by lower saturation and lighter hues (yellow, blue, light brown, pale green). Where the two requirements of 'natural' appearance or association and desired visual level coincide (as in the brown used for contours, blue for water, and green for vegetation), conventions are firmly established. They are firm because they work successfully both in the sense of characterization and in the sense of the required level of graphic emphasis. In the classic problem of hypsometric colours, the unavoidable association of colour with the surface characteristics of areas, and the effect of elevation on climate and vegetation, does not correspond to the need to produce a long scale of discriminable values.

So that although some organizations tend to follow a standard practice, there is no solution which can be regarded as solving the representational problem sufficiently well to establish a convention.

The combination of hypsometric colouring with hill shading demonstrates the contrasting uses of gradations of area colour. The hypsometric colours are denotational: they are intended to emphasize differences in elevation, even though they cannot avoid some association with landscape characteristics. If hill shading is superimposed, then this is intended to be naturalistic, to give a direct visual impression of relief. Unless the relative effects of the two types of symbolization are carefully controlled, they are always likely to result in conflict.

Reeves (1910) was one of the first to appreciate the connections between landscape painting and cartographic methods. With regard to aerial perspective in relation to hypsometric colouring, he says, 'You will have noticed that it is much in this way that a painter gives the effect of nearness and distance to his pictures. In the foreground he introduces reds and yellows, and then for the distances, blues and greys in combination.' Unfortunately, in this case, the requirements of aerial perspective are not in sympathy with either the colour associations of natural landscapes or the presence of other information on maps.

The naturalistic use of colour has both an emotive and a representational effect. A photomap of part of the Arabian peninsula, printed in an orange-brown, not only gives a suitable background image in terms of contrast with other elements, but also reinforces the impression of dryness and lack of vegetation.

Colour knowledge and colour terms

Despite the difficulties which painters encounter in handling their mediums, the practical experience of selecting, reducing and mixing colours on the palette is closely connected with their efforts to achieve particular effects. Painters rarely use pure hues in representation – they are usually too stark and vivid. Where they occur, they are more likely to be applied in small areas, or at high contrast points. In creating a picture, the painter frequently makes complex colour combinations in order to obtain the desired effect. Adding white to lighten, or a touch of grey to darken, are common practices. In addition, the distinction between the many possible variations in single and mixed colours is reflected in the painter's vocabulary. She does not refer in simplistic terms to 'red', but much more specifically to vermilion, crimson, red ochre, red lake, scarlet lake, carmine, and rose madder. It is evident that the source of these distinctions lay originally in the nature of the different natural materials used to produce pigments, for these were the materials with which she had to work. Indeed, many of them were further distinguished by their area of origin. For example, Chinese vermilion was not the same as French vermilion. This practice in dealing with colour and its creation focuses attention on the great range of possibilities available graphically. Even if learned empirically, rather than theoretically, a knowledge of colour is an essential part of the painter's preparation.

The numerous publications on painting produced in the later nineteenth century demonstrate an interesting preoccupation with colour. A more widespread interest in colour, especially in reproductions in colour, was certainly one consequence of the rapid improvement of lithographic printing, and the development of the half-tone process at the end of the century. Part of this interest was scientific and technical. Spectral analysis, the colour theories of Young and Helmholtz, a widespread understanding of primary and secondary colours, colour mixing, etc., were linked to the development of pigments, vehicles, varnishes and a considerable industry of a technical manufacturing nature. Works such as those by Collier (1890), Vibert (1892) and Parkhurst (1898) were largely technical manuals, dealing extensively with materials and methods.

Colour-separated production

In this respect cartographic production in the colour-separated mode is graphically unfortunate, however effective it may be in terms of technical production. The problem is that few cartography students have acquired the familiarity with the use of colour, especially the effect of colours in combination, which artistic practice would have given them; this is compounded by the fact that during the analogue map production phase the images are normally achromatic. Their full representation in colour, and the effect of relative lightness and darkness, has to be imagined. All too often, the first colour proof comes as a considerable surprise to the novice cartographer. Unfortunately, the type of manuscript sketch that can be produced quickly is usually a poor match for the types of image actually created by modern production methods. Perhaps the greater availability of sophisticated digital cartographic systems will make possible more opportunities to experiment with colour design and combination. Yet even this is unlikely to be realized unless the cartographer can conceive the potential effect of possible design solutions. This requires an understanding of colour, not just a technical means of controlling it, because a file of digital data has to be assigned a specific colour before it is presented on the screen as part of a map, even if this is done provisionally or even accidentally.

The further consequence of this lack of acquaintance with colour is less obvious but just as important. In map production, the printing colours are normally determined in advance. Therefore it is technically much easier to use these hues without modification. Consequently, the cartographer tends to think in terms of these predetermined colours, rather than those which would be more appropriate to the particular work. By identifying colours initially in simplistic terms, such as red, blue and yellow, or cyan, magenta and yellow, the opportunity for greater elaboration of the coloured image remains unconsidered.

The fact that in many cases the available colour palette for a printed map is limited makes the problem even greater. It also induces the danger of conceiving a coloured map as essentially a series of coloured lines or a set of outlined areas to which surface colours are added, and in which colour is thought of simply in a denotative sense. The difficulty of avoiding this is evident in a large proportion of the multicolour maps produced in Europe, especially in the first half of this century.

Unfortunately, the tendency of modern cartographic practice, in particular the trend to standardized printing colours, frequently works against a proper consideration of the possibilities of colour. Although advanced under the banner of efficiency and economy, any technical procedure which limits graphic choice works against the achievement of a good composition. It is rather like asking a dramatist to write a play with an arbitrarily restricted vocabulary. A painter will labour in the production of a particular hue or tone, manipulating pigments on the palette until the desired colour is achieved. In dealing with area colour, quite often the addition of a touch of grey achieves a slight desaturation of the hue, and alters the effect. To achieve this in cartographic practice, when all modifications to the printing colours have to be technically produced, is both difficult and time consuming. Yet in so many cases the cartographer is forced to accept a set of standardized hues as though this made no difference. Frequently the control or adjustment of a composition depends on fine variations in colour and shading, and not on the replacement of one symbol by another which is totally different. Even in a coloured composition, it is the relative lightness and darkness of different elements in the image that is critical, not simply their hues.

The importance of this control over the total image can be seen in the medium-scale topographic maps of European countries. What the most successful – both past and present – have in common, is a careful and deliberate choice of primary colours, as well as their combination and manipulation. It is the balance between them that is so critical. As Craig (1821) points out, 'the painter, to ensure a successful combination, must have his eye made accurately susceptible of small differences and agreements.'

Process colours

The emphasis on the so-called process colours as a means of reproducing in print any coloured image rests upon a fundamental misconception when applied to coloured maps. With colour separation, the process colours can be used to reproduce any type of coloured image, within the limits of lithographic printing: they are not used to create the original image in the arts, whether produced by painting or colour photography. Therefore they do not restrict the possibilities available to the artist or photographer. When determined in advance for map production, they tend to restrict the types of coloured image that will be produced; not because any colour mixture is technically impossible, but because specific colour mixtures become technically complex. Although such colour mixtures are familiar in the symbolization of area colours, they are rarely attempted for the line image, with the consequence that many possibilities are not even considered by the cartographer. The fact that colour charts can be used to show the range of solid and tint combinations possible for area colours does not solve the visualization problem, for these rectangles do not indicate the effect of a given colour when applied over large or small areas, or how it will appear in combination with other colours. They certainly give no visual impression of how a mixture of the subtractive primaries will appear if used in combination in fine lines or over small areas. Thus the imaginative use of colour in the context of the composition as a whole is inhibited. The effect is insidious, for many cartographers appear to be quite willing to work within the confines of notional terms such as cyan, magenta and yellow, or red, blue and yellow, apparently unaware of the possibilities of selecting specific hues within the compass of these very general terms.

One quotation from a nineteenth-century book (Vibert, 1892) will demonstrate the point. Although this is largely a technical manual on colour pigments, a section on colour contrast states that 'When two colours which follow each other in the order of the spectrum are placed side by side, they take more and more, in proportion as they approach each other, the aspect of the colour which precedes or follows them. . . . Yellow beside green becomes more orange; green beside yellow more blue; blue beside ultramarine becomes greener.' This understanding of colour arises naturally as the artist mixes and works with pigments directly. To achieve the same level of understanding with images constructed separately and in the achromatic mode, as happens in most cartographic production, requires a great effort of will, as well as imagination. In the recent past, the overwhelming insistence on the technology of colour reproduction and its unthinking application to cartography in itself demonstrates the characteristic preoccupation with economic efficiency and the neglect of art.

Colour in digital map production

In the present period many digitally produced maps follow the design characteristics of their prototypes. But if the advanced systems coming into use can enable the operator to choose any colour from a full range, presenting it on the screen in relation to the other coloured symbols, it would be possible to free design from the restrictions of the process colour approach. Such a development presupposes that the analysis of the component primary hues would be done subsequently by the system, and that the printing of any hard copy would be able to achieve the resolution and register needed for a high-quality line image. But even if the technical problems could be completely solved, the effective use of such a system would still require cartographers who could think about and imagine colours in the widest possible sense.

Although digital colour separation is currently used in the reproduction of existing images, that is quite different to the generation of a new design. Technically, such a system would be highly demanding, but if it could be implemented economically, it would represent a major breakthrough in cartographic design. Such a development would indeed merit the description 'revolutionary'.

Art and education

What both cartographers and artists have in common is the need to control the graphic medium in order to make expressive representations. In art education the need to practise and experiment is balanced with a constant attention to the study of existing works. In cartographic education, technical knowledge tends to predominate. Because there are so many other things to learn, spending time simply looking critically at maps tends to be neglected. Yet it is only through the combination of practice and visual attention as to how effects are achieved that the necessary skills can be attained.

This is partly a consequence of a general ignorance of the graphic arts in society as a whole, and this affects map users as well as cartographers. Although some instruction in drawing is provided in most schools up to a certain age, the wider field of looking at, appreciating and even criticizing a variety of art works does not have some measurable or assessable outcome. What is missing in most cartography students is not lack of intellectual ability, or enthusiasm, but a broad awareness of graphic art and the acquired habit of visual attention.

The same is true in relation to map users. Whatever their academic attainment, using their visual perception to acquire information tends to remain at the most rudimentary level. The difficulty that many have in map interpretation is not just lack of knowledge about maps, but lack of any development of visual perception that would have taught them to react to visual images.

The lack of 'graphicacy' has been lamented by many writers concerned with map interpretation. It is doubly unfortunate because children in general display a natural interest in drawing and painting which is allowed to wither as soon as 'serious' education commences.

In reviewing the development of teaching art in schools, Carline (1968: 4–5) distinguishes between the teaching of art to intending artists and the role of art education in general. Both are of significance in relation to cartographers and map users. Making the point that 'Art education is advancing too slowly and its place in the life of the community is far behind many other activities which are less essential to human well-being,' Carline maintains that 'The school art class should not only develop visual sensitivity but provide a background of taste, perception and appreciation which can be helpful throughout the school in studying a variety of subjects as well as art.' Despite the remarkable progress in the teaching of art to young children, even gifted pupils are likely to be 'withdrawn from their art studies because they have revealed an equal promise in some other branch of study.' The consequence is 'liable to lead us further along the road which ends in a divided society, with those who think, see and behave as artists – the minority – on the one side, and those quite unconcerned with art – the great majority – on the other side, both sides eyeing one another with suspicion.'

Many young people who set about the study of cartography, attracted to it out of what is often a subconscious interest, are often ignorant of the most elementary terms in the graphic arts. In many cases, they regard the artistic element in cartography as difficult or incomprehensible. Thus they look for rules that will achieve what they sense but do not understand. In the same way, any attempt to talk to map users about artistic elements in maps immediately raises difficulties in comprehension. The idea that art is not 'serious' or worthy of attention is overwhelming. The desire to find 'scientific' solutions to problems of map design emanates from deeper roots than theories of communication. The refusal to recognize the need to acquire artistic skill in the full sense, as well as the reluctance to accept that the artistic dimensions of cartography require serious study, affects the capacity of map users to appreciate and interpret maps just as much as the ability of cartographers to create them.

CHAPTER 16

Generalization and composition

At first consideration, generalization and composition may seem unconnected. Certainly their relative importance in painting and cartography is different, although both are present. Whereas composition is a basic topic in painting, the term is rarely used with regard to maps. Conversely, whereas generalization is a fundamental subject in cartography, it seems to be less apparent in painting.

Yet it is interesting that many of the key terms used in one appear in the other. Detail, simplification, importance, character and balance are all widely used. Thus the connections between the two are worth exploration.

Generalization

In virtually all derived maps, questions of generalization are bound to have arisen, whether or not they have been dealt with satisfactorily. It should be stressed (because of the widespread misuse of the term 'design') that the adjustment of the information to the scale by generalization is primarily a compositional problem, not a design problem. It is necessary to say primarily, because of course the design of specific symbols will also influence the composition and generalization. The width and colour of a contour line may be decided at the design stage, but they also influence the chosen vertical interval and will thus affect any generalization. As with all artistic works, everything in the visual appearance of the map is interconnected.

Given that a basic topographic map should be provided with the information required for a given scale consistently, the level of generalization itself will be chiefly a function of scale. The more diverse the source material – and in many small-scale derived maps or special-ized maps it is likely to be diverse – the more complex the question of generalization becomes. Given that the effects of generalization are most apparent and most obtrusive with small-scale maps, the whole subject is of great significance in small-scale cartography.

The question of scale seems to be less apparent with painting, yet it is obviously an important variable. The size of the representation itself, whether large or small (equivalent to the map format) is one factor. Extensive views will necessarily involve considerable generalization. The choice of dominant subject, whether some object in the foreground or a distant view, also affects the detail of treatment. The degree of detail given is partly a function of the size of the picture (the overall nominal scale), and partly a choice of dominant subject. If the chief subject, for example, is a building in the foreground, then this will receive a more detailed representation than the same building if it appeared in the distance. This is itself partly a function of perspective, and partly a choice by the artist in terms of the subject matter of the picture. In more formal arrangements, a still life is usually treated in a detailed close up, at a relatively large scale. Therefore the degree of detail and generalization in a picture is a function of subject matter and composition, whereas in a map it is primarily a question of the purpose of the map and the overall map scale.

Selective omission of detail, simplification of form and the combination of small individual elements into masses are typical of representational drawing and painting. Areas of woodland in the middle distance or background are shown as masses; distant buildings have highly simplified shapes. Except in close-up views, foliage has to be treated as patterns or masses of light and shade. The simple landscape sketch shown in Fig. 42 illustrates these points. According to Gilbert (1885),

> Generalization avoids or simplifies certain complexities of form, because, if these be laboriously detailed, they would encumber that which is its most excellent feature. Generalization, no doubt, omits certain facts, but it omits them only that it may more vividly present the fact of greatest interest. . . . In this way is given the larger truth, for literality defeats itself; attempting too much, it reaches less.

Trying to put in too much detail is a common failure with the student, whether of painting or cartography. With maps, it is most apparent with lines and outlines, which lend themselves to a level of detail which is frequently out of keeping with the scale, and serves no purpose. At the same time, indifferent line simplification, which loses the character of the feature, is just as injurious in painting as in cartography.

In a picture based on central perspective, the relative sizes of objects, and therefore the detail in which they are treated, is a function of distance from the assumed viewpoint. Only those objects or features which are close will be large enough to be given a detailed treatment. But the illusion of depth is reinforced by using higher contrast in the foreground, and subdued colours in the background. Quite often, a detailed treatment of an object close to the viewpoint, such as a tree, is deliberately introduced to help in the required visual contrast and perception of depth.

With a map the scale is nominally constant. Although with a

Figure 42. Generalization in landscape drawing

large-scale map this nominal scale may apply fundamentally to the whole of the content, there will always be exceptions, as small features or abstractions have to be represented by legible symbols. For medium- and small-scale topographic maps the requirements of legibility and relative importance will introduce sufficient exaggeration to ensure that important small features will be retained, regardless of nominal map scale. In this sense they remain part of the 'foreground' in pictorial terms. So in both cases scale and generalization are used in combination, although to different ends. By choice of composition, the picture ensures that the primary subject matter is brought into focus by using the foreground. By manipulating selection and symbolization, the map ensures that what is regarded as most important is included and if necessary emphasized regardless of absolute size, and so given detailed treatment, within absolute scale limitations.

With maps of special subjects, which generally combine two sets of information, the particular subject and the reference base, the subject of generalization is just as complex, although it may be less apparent. The scale of the map may be that which is required, chosen so that it is large enough to present the detail of the information, just as the artist may choose a large enough canvas to deal with a particular scene in accordance with his intentions. On the other hand, the scale of the map may be enforced, i.e. the level of detail present in the subject matter may have to be condensed or reduced to fit within a given scale and format. In this case the degree of generalization will be increased, just as in a small painting. In these circumstances the painter will concentrate on what is visually important, whereas the cartographer must concentrate on what is informationally important. Scale and content set the limits to the possible. Both artist and cartographer use the graphic variables within these limitations to achieve the desired end.

For many maps based on statistics, the question of generalization is often neglected. Yet it is crucial to small-scale maps of this type in particular. Devising a range of proportional symbols, for example, that would reflect in detail the variations in quantity or value of a given subject, could often be achieved graphically if the map was large enough. This would allow space for a large number of differently sized symbols to be legibly represented, without crowding one on top of another, and at the same time permit a greater range of absolute symbol size. Yet the majority of such maps are so small in scale that the problem of representation becomes very difficult to solve. Although this type of map has been extensively discussed in relation to choice of method of symbolization, limitations of visual discrimination, etc., it is itself largely a generalization problem, enforced by scale. It is rather like asking a painter to produce a large landscape in a miniature, and then complaining that it does not show enough detail, or is 'difficult to see'.

An understanding of the effects of generalization is just as important to the map user as it is to the cartographer. The generalization processes of combination, exaggeration and displacement, which are normally applied to achieve either symbol legibility, or content emphasis, or both, are most obvious when considered in relation to tangible features on the Earth's surface, such as buildings in urban areas. Yet graphically they apply equally to other kinds of phenomena. A map of population distribution in a given area, using actual population numbers as the information base, and employing the 'dot' method, is just as much a problem of generalization as any other kind of subject. If constructed at a small scale, which is quite common, not only will the topographic outline be subject to selective omission and simplification, but the representation of aggregates of population by legible 'dots' will involve selective omission of thinly populated areas, combination into population 'clusters', exaggeration of symbol size, and quite often lateral displacement. In this respect the way in which the map is composed will be intimately linked with the degree of generalization.

Composition

Composition is a term that is used in relation to all the creative arts, from music to photography. According to the *Shorter Oxford English Dictionary*, to compose is 'To make by putting together parts or elements; to make up, frame, fashion, produce'; and composition is either the act of combining or forming, or the product itself, such as the practice of literary production. In the rhetorical sense it includes both *inventio* and *dispositio*; the choice of subject matter and its disposition or arrangement must be considered together.

As is often the case, the term has several different connotations. With music, composition can refer to the whole of creative work of the composer, and 'a composition' is the complete work. In the case of painting, it is usually applied in a more restrictive sense, to describe the process of creating the overall structure of the work, laying out the various parts, but does not include the final treatment of detail.

Laidlay (1911) quotes Sir Joshua Reynolds as saying that

> Composition taken generally is the principal part of invention, and is by far the greatest difficulty an artist has to encounter. Any man that can paint at all can execute individual parts; but to keep these parts in due subordination, as relative to the whole, requires a comprehensive view of art, that more strongly implies genius than perhaps any other quality whatever.

According to Laidlay, composition includes, in its widest sense, 'a consideration not only of the lines in a picture, and its light and shade, but also a careful study of what is known as "balance", and "harmony", and above all, a thorough knowledge of the arrangement of masses of colour, when in union or opposition.'

The notion of composition operates differently in different types of painting. For the religious and usually allegorical paintings of the medieval and Renaissance periods in Europe, composition was employed to construct imaginary representations of historical or religious events, centred mainly on groups of figures. These pictures are generally analysed in terms of their spatial organization, leading lines, and so on. On the other hand, composition in landscape painting starts with a chosen subject. The way that this is 'framed', the viewpoint, the possible choice of a near object or a distant one as the centre of attention or focal point, and the absolute size of the picture and its shape are all elements in composition, as well as the relative importance of its parts or elements. In many cases the type of composition reflects the style of the artist. For example, with many landscapes by Turner the effect of the sky and its treatment are an important element.

Cartographic composition

It is difficult to understand why this topic is so neglected in writing on cartography. There seems to be an assumption that once the scale, area and content of the map have been determined, and as the subject matter cannot be 'rearranged' because of the need to keep correct location, that 'design' exists quite separately from composition. It is a notion that would surprise any artist. Paintings produced by commission, and paintings or frescoes produced to occupy specific spaces in religious buildings – and it is possible to think of many in this category – certainly had a predetermined subject, but the way in which the subject was 'arranged' and treated was a fundamental part of the work.

It is one of the characteristics of many types of map production that the circumstances in which maps are initiated, and the decisions taken about major aspects of content, scale, area and format, are not necessarily considered as a whole. The 'cartographer' often tends to see the composition of the map as already given, as decreed by external factors. Quite commonly, the division between the provision of information and its cartographic representation ensures that in many cases consideration of design is subsequent to the conception of the map as a body of 'information'. So the 'survey' or the selection of a set of 'data' takes place first, and the problems and issues which arise in its cartographic representation are treated at a secondary level. In very few cases is it obvious that the map as a whole has been thought through.

The basic characteristics of a map as a composition are determined by the scale, the geographical area, the intended information level and the format of the map or map sheet, all of which are interconnected (see Keates, 1989). Whereas arrangement in the sense of placing things spatially is predetermined by geographical fact, the choice of content is not automatic, but a matter of judgement. But arrangement in space, and in particular the density of different phenomena on or in relation to the Earth's surface, exercises a considerable influence on the choice of both scale and content. It is the fundamental reason why the definition of any map must include its scale, for scale is a conditioning factor in both content and its treatment. In this sense it is impossible to conceive of just 'a map' of any area. It is meaningless without an indication of scale, for this immediately affects the informational content.

Therefore the decision to produce a map of a given area and subject matter, at a particular scale, and within a given format, must be the equivalent of the first stage in composition in painting. In compositions that are essentially linear, as in both music and literature, this framework may be rudimentary; it may expand or diminish in its different parts as themes are explored and developed. In this respect, the initial plan or concept may be only notional, for different parts can be retrospectively developed. In any two-dimensional work, this is not the case. Given a certain basic scene or view in a landscape painting, the choice of height and direction of viewpoint, the relative proportions of different objects in foreground and middle distance and so on, cannot be changed arbitrarily. To do so requires a different composition, and therefore a different picture.

With maps, this sense of composition is all too frequently missing. Yet it must be the experience of any practising cartographer working for a map author that the initial problem is to bring the area, scale, information and format into harmony, because cartographically they have to be treated as a whole. In many cases, the notional scale first suggested is quite unsatisfactory for the level of information desired. Either one or the other must be changed. A selected geographical area which has to be 'fitted into' a fixed format makes possible a particular scale, and therefore limits the maximum density of information. Awareness of this is an essential part of a cartographer's knowledge when discussing proposed maps with clients. Although sometimes referred to as the 'planning' stage, it has to deal with particulars, not abstractions.

Lack of attention to composition is just as dangerous in cartography as in any other art. Maps in which the scale is obviously too large for the content, and maps in which the content is too detailed for the scale are the obvious examples. Presented with a body of 'information', the cartographer must be just as aware as the landscape painter of the connections between content and its representation, for both depend on visual imagination based on understanding. In both cases, the sense of what is possible, what can be graphically realized, is derived largely from experience. But for an artist this experience is deliberately sought, by examining and studying existing works, and by making 'studies' of subjects and their composition before embarking on the full work. The need to study maps, not as sources of information, but in terms of their merits as compositions, ought to be a significant part of any instruction in cartography. Just as the artist can study composition in many different types of picture, so the cartographer should also be conscious of the compositional merits of different maps.

With a landscape painting, the subject itself can be freely chosen by the artist, who is under no obligation to produce information about it. The treatment is of course rhetorical in the sense that any painting is only one of innumerable possible creative interpretations. The same painter may well paint the same subject in different ways. It would seem therefore that the map, for which the content itself has to be composed, must be very different. The painter can at least start by viewing the complete scene; his conception of the possible representation, though not a copy, at least starts with an end in view. On the other hand, the map maker has to start by selecting information about individual things, and then composing these into a whole.

Despite these differences, both painter and cartographer have subject matter to deal with. Once the painter has chosen the particular scene, he then has the task of creating something based on it, however loosely. Once the subject, area and scale of the map have been determined, the cartographer is also faced with given content which has to be visualized and presented. Just as a painter can produce his representation either in the manner of his individual style, knowing in advance what a certain treatment is likely to achieve, so the cartographer can operate on the basis of established conventions and personal experience. How to use line weight, symbol size and contrast between fully saturated hues and pale tints is just as embedded in the cartographer's subconscious as in the painter's.

Only in those cases where the particular map is simply one more in an established sequence of familiar maps is the composition likely to proceed along standardized lines. Even when what appears to be suitable information for a new map has been obtained, decisions about the scale and the relative importance of different aspects of the content have to be taken in advance, before the design of the map can begin. But this provisional arrangement may be only the starting point. If it proves to be impossible to deal with the presentation of the information satisfactorily on the basis of the initial composition, then either the information or the composition will have to be adjusted. If the map is sufficiently simple, this consideration may take place only in imagination. But it still has to take place.

For both artist and cartographer, the formulation of the composition will overlap with problems of design and representation. The experienced cartographer will mentally test possible compositions, imagining the effect of particular symbols, background colours, lettering sizes and so on. The important connection here is that the experienced cartographer, like the painter, will mentally introduce and adjust various combinations to form possible solutions, considering the effect of subduing a background feature or emphasizing a foreground feature. Much of this is done by visual imagination, just as the painter imagines the effect of certain forms and colours.

It is in this sense that an unsatisfactory composition, in which there is some mismatch between the chosen scale and the intended content, will be more quickly realized by the experienced cartographer than by the beginner. Often the beginner will struggle to find some satisfactory solution to what seems to be a design problem, without realizing that the basic composition is at fault. It is commonly the case when the scale of the map is too large for the content and informational level. Although all the 'information' is there, the map seems to lack coherence. Yet if the scale is reduced, the map is made more compact and requires less visual search.

For both maps and pictures, the ultimate composition and design may not be initially apparent: for the artist, because the nature of his eventual interpretation may not be fully evident at the beginning; for the cartographer, because the implications of the composition and the consequences of representational methods may not be apparent at first study. In some cases the composition itself may have to be changed or modified. The significance of a certain level of importance for one element in the whole may not be realized until the image has been partly constructed. Whereas the artist will do this directly in the course of painting, it is more difficult for the cartographer to achieve the same realization without actually making the map according to a given concept. In both cases the initial basis may be discarded, or considerably modified. In extreme cases, the painter will simply abandon the work and start again. With the map, the initial composition and design may prove to be unworkable, in which case either the initial composition has to be abandoned, or the content itself changed. The painter who realizes that he simply is not getting the effect he wants may have to reconsider either his selected scene, or its treatment. The cartographer who cannot reconcile the desired content, scale and representation may also have to change the content, or the representation, or both.

It is common for artists to make several exploratory studies or sketches, either of the whole scene or selected aspects. Unfortunately, because it is very difficult to 'sketch' in a manner resembling a fine-line printed image, there are fewer possibilities for quick experimental works by the cartographer. In addition, in most commercial mapping operations, the client is unlikely to allow part of the budget to be reserved for experimental designs. As Petchenik (1983: 65) points out, 'It takes time and money to carry out research that would show whether a particular map was significantly flawed and, generally speaking, in the commercial sector there is little or no desire to allocate resources to such research. Most maps could not be re-made even if they were found to contain design weaknesses.' And she continues, 'Many maps made in the commercial world are for one-time-only publication, thus providing little or no opportunity for alteration in the light of evaluative research or user response. The permanency of our design decisions may be intimidating, but it is a fact of commercial life.'

It is worth repeating that the ability to generate and change images in a visual display is one of the great potential advantages of digital mapping systems. But it will still be necessary for the cartographer to imagine and visualize the effects of possible changes in order to gain from the system's speed of operation. Solutions will not present themselves by a random pressing of keys, or changing one feature without considering the effect on all the others.

Composition and the rhetorical model

The rhetorical model emphasizes the interconnections between content, representation and design. As such it coincides with the reality of cartographic practice much more than any communication model that stresses external aims or conditions. It also counterbalances any rule-seeking research, because it is appraisive, not prescriptive, and concerned with the whole product and not just individual parts.

A preoccupation with representational methods does not necessarily lead to a satisfactory result. Its consequences are most evident in maps that have been constructed by 'borrowing' apparently successful methods from a variety of sources. Once divorced from their original compositions, they may no longer provide the desired effect. In a medium-scale topographic map, a method for line simplification may operate successfully for an individual feature. But a set of closely spaced contours along a slope are not just a set of lines, they represent a surface, and as such, choice of vertical interval, line gauge and colour will all influence the generalization. They also have to be treated in relation to the drainage system and the selection of retained spot heights. The representation of relief, of which the contour lines are a part, has to be considered as a whole, not as a set of isolated actions.

The same interconnections are evident in name arrangement, which is also a compositional task rather than just 'design'. With many small-scale maps, the selection of towns and cities to be included in the content is in turn affected by the possibility of actually inserting their names, which is why the assembly of the content and the name arrangement have to be worked together.

It is commonly the case that the production of a new map, in its initial stage, reveals some unsatisfactory element. Chance combinations of certain symbols may result in one of them being illegible in some part of the map. An initial design specification may mean that some feature is over-emphasized. Reacting to these is as instinctive for cartographers as it is for other artists, for it is the relationship between all the individual parts that poses the most taxing problems. Refinement of these details can only arise as the cartographer responds to the map as a complete work, and it is this visual response that is critical. As with any artistic work, a 'good' map is one in which the content, scale and representation are treated in harmony, and therefore contribute to the work as a whole.

CHAPTER 17

Style, fashion and taste

Discussion of both works of art and artifacts frequently refer to them in terms of their style. This may be applied to the analysis of the qualities and characteristic of an individual work, in order to describe and evaluate it, or it may be used as a means of placing different works in recognizable groups which have common characteristics. This is apparent with pottery, furniture, clothes ard even motor cars, as well as the more durable components of a culture such as buildings. In literary criticism, the question of style is regarded as important, and is used as a means of contrasting and even judging literary works. In painting, virtually all individual products are categorized by type, taking into account such things as subject matter, medium and the individual manner or method of expression of the artist.

With this in mind, it is strange that so little reference is made to style with regard to maps. Historical accounts deal with some of the ingredients, such as the biographical details of individual map makers, the circumstances in which they worked, the influence of existing products, the effect of new ideas, the introduction of different methods, and so on. Yet they rarely seem to proceed to any conclusions that would attempt to summarize either individual maps or groups of maps in the stylistic sense. It is even more strange when the efforts of many map and atlas producers to make their products distinctive and individually recognizable are taken into account. It is also clear that at least in some ways maps follow fashions, and fashions themselves are deliberate attempts to create and popularize particular styles.

For the perceiver, user or consumer, questions of style are also bound up with taste, with the way in which appearance or form affects the interest and judgement of the user. Therefore it is intimately connected with aesthetics, and in some cases it is the superficial appearance of the object which becomes the focus of attention, rather than any fundamental quality. Thus the regular stylistic or 'cosmetic' changes of the motor car industry serve to advance both the concept of newness or creativity (a new model), the aesthetic appearance of the product (sleek lines or a low profile) and current fashion (being up-to-date), as well as the introduction of true technical improvements. On the other hand, the continuation of a particular form in terms of body shape or exterior finish serves to maintain the individual style of the products of prestige makers. Again, the care with which motor car manufacturers gradually introduce any fundamental or major changes in appearance indicates the innate conservatism of the general public, whatever the enthusiasm of individuals. It has its counterpart in the changes introduced as a new edition or map series replaces an old one. In this sense, taste can be regarded as either whatever is popularly acceptable at any time, or the special capacity of those who have a highly developed interest in and understanding of the arts.

The analysis of style

The factors which should be taken into account in describing style have been thoroughly analysed by Munro (1946). He draws attention to the vagueness and ambiguity of many general terms, which frequently confuse historical with abstract factors, such as baroque or romantic. It is necessary to distinguish between the concept of style as the character-istics of an individual artist, and style as a combination of characteristics that can be used to classify or describe many works in a given period or region. Both are related to the presence of characteristic traits, elements of composition, subject matter, use of terms, representational methods, imagery and so on, which give particular works a definable and different character.

Munro identifies the use of historical periods (e.g. Ming dynasty), geographical regions and people (e.g. Japanese prints, English water colours) and abstract traits used in a summary sense (e.g. Impressionist, Cubist). He points out that these traits may involve the material, medium and technique; the treatment of components; and modes of composition.

Schaper (1972: 644) also distinguishes between period derivative names, feature names, non-derivative feature names and normative style names. The first group includes direct reference to historical periods, such as Louis XIV. Feature names include such groups as early Florentine, and are frequently geographical in origin. Medieval maps and English County maps (generally referring to those produced in the seventeenth and eighteenth centuries) are possibly the closest to these two groups. Non-derivative feature names are based mainly on formal or expressive elements, such as Gothic. Normative style names are themselves based on normative standards, such as classical. In Schaper's view, 'A style name always indicates something collective, something by virtue of which one work is comparable to others.' A collective style usually develops as a workable solution to a given type of subject matter that can become a recognized way of doing, and therefore a convention. Many modern topographic map series fall into a stylistic group, in the sense that they are based on an accepted and effective method of design and representation. What is called the 'Swiss manner' is probably the best example.

Style in cartography

From these brief considerations it is evident that any discussion of style in cartography must take into account two related aspects: the particular qualities or characteristics of individual maps, and the possible existence of schools or periods which have distinguishing properties. In both cases several difficulties immediately become apparent. Works of art are usually associated with individuals, or groups led by individuals. Even where they are known collectively by the name of the place or manufacturer, as is often the case with pottery, it is usually possible to trace the individuals who carried out the production.

In earlier times, maps tended to be associated with individuals, although it is sometimes difficult to tell if an individual map was produced by the named cartographer, or by one of his apprentices. For example, Steer (1967: 125) refers to 'the perpetuation by a younger man of certain styles, either of the map itself, its lettering or decoration, which may have been the peculiarity of an older man.' In the modern period, with the replacement of individual craft by industrial manufacture, the separation of design from production and the need to employ numbers of people in the production of a particular map have made this close association of individuals and products generally impossible. In this respect, most maps are products of agencies, in which the individuals are rarely named. Some maps originate in small companies rather like artists' *ateliers*, in which the ideas and influence of a leading individual are known. However, in most cases the cartographic input remains anonymous, even though it must have been present.

In another sense this reflects the degree of importance given to cartographic design, and the role of art in cartography. It is clear that in the nineteenth century, for example, the makers of topographic maps generally followed whatever had become the style or manner of their particular organizations, assuming this to be the normal or proper means of expression, and giving it little attention. Most of the elements in the style of the topographic maps of any one country resulted from an evolution of practice into conventions. Changes in style were usually a consequence of new content requirements or changes in the technical medium. A great deal of topographic drawing simply followed what seemed to be the manner of the time, treating it as a production process rather than a creative one. This emphasis on production was important, because production was easier and faster if people followed the practices which were familiar, and output was the primary aim of most mapping agencies. Changes in representational methods were welcomed if they made production more economic, and resisted if they required more time, skill or effort. In this respect production needs exert a pressure on artistic expression. In the modern period, pressure to produce maps more economically has been a major factor in the appearance of many topographic maps. It is evident, for example, in the conversion of the British 1 : 63 360 Seventh Series, originally designed in ten colours, later reduced to six, to production in the four process colours for the first version of the 1 : 50 000 scale map. Since the early 1980s the drive for faster production and greater economy has also affected all the United States topographic map series, not only short-cutting such things as field completion, but also in modifying the graphic design toward more simple forms.

Although the artistic qualities of such maps were accepted, they were generally connected with the concept of art as skill, rather than art as creativity. Even when a new design in the full sense was introduced, such as the Fifth Series of the Ordnance Survey One Inch to One Mile Map, which was radically different in style to its predecessors, there appears to be no published documentary evidence that explains exactly how this new product was devised, or who was responsible.

The cartographic medium

Although graphic art works are frequently classified by the medium used, such as oil painting, water colour, lithographic print, etc., this is only one element in analysis. For maps it has limited use, mainly because one medium has tended to dominate a given time period. With the introduction of engraving, virtually all maps for a long period were engraved sheets, and therefore shared a common medium. Although there was a considerable overlap between the disappearance of engraving and the introduction of lithography, eventually lithographic printing became the dominant medium – a position which it has maintained up to the present.

Different mediums certainly affect the degree of uniformity or variety of individual maps. It was characteristic of engraving that the medium itself exercised a strong influence on what was graphically possible. It depended heavily on delineation, and if colour was included this could only be done by subsequent addition to a line image. Modern lithographic colour printing, on the other hand, offers a full range of possibilities of both colour and tone, and this freedom of graphic expression has had a marked effect on the content, methods of representation and aesthetic appearance of printed maps. In other cases, the medium is specifically limited, normally because of production costs, so that maps produced in a limited number of printed colours are restricted in what can be attempted, and therefore in the range of graphic expression. The development of digital cartography, and the variety of image-forming systems that have been developed in response to digital mapping requirements, have led to the beginning of a new phase. Its influence is already apparent in the replacement of the double-line road symbol by a single line, sometimes combining solids and tints, and the reduction in the number of interrupted line symbols (Jannace and Ogrosky, 1987). Both of these changes reduce the range of graphic expression in the interests of more economical production.

At present it is by no means clear as to whether any single medium will dominate in the future, or whether several different mediums will co-exist. Certainly the production of a map as a visual image on a screen may have a definite effect on what can be accomplished in a technical sense, for the projected image viewed by transmission has different visual properties to an image viewed by reflection. In this respect, map making is going through a period of change, the consequences of which are not yet predictable.

Functional classifications

It is also possible to classify art works by type of subject matter, such as portrait, still life, landscape, etc. Such a classification runs across that of medium, for the same subject matter can be expressed in different mediums in this respect, landscapes may be in oils, water colour, line and wash, or even line only. The nearest equivalent cartographically is presumably the major division into topographic and thematic, although this classification is inadequate. Specialized maps, as opposed to general ones, may deal primarily with a particular subject, or a particular purpose, and the distinction is important. Such a classification is by far the most common in the general description of map type. Its most obvious use in the English language is the distinction between map and chart.

The other element that is significant with regard to maps, but plays little part in other artistic works, is that of scale. Although specialized cases are distinguished in the fine arts, such as miniatures or murals, scale is not normally regarded as a principal factor. With maps, scale is highly significant, and scale differences are inextricably linked to both map type and representational methods. Indeed, it is generally impossible to make any comparison between different maps if they are at appreciably different scales, because of the effect this has on content and classification. Even so, the products of any one organization may retain a particular style throughout a wide scale range, as, for example, in the United States 1 : 24 000, 1 : 62 500 and 1 : 100 000 scale series; the Swiss 1 : 25 000, 1 : 50 000 and 1 : 100 000 series; and their equivalents in other countries such as France, the Netherlands and Denmark.

Topographic maps

Topographic maps provide an interesting source for consideration of stylistic differences mainly because they are so widely produced at a range of scales, and because they have existed and developed over a long period. As 'general' maps they have ostensibly a common character in terms of subject matter. They are dominated, in terms of published sheets, by the output of major national agencies. The degree of available documentation about their provenance is extremely variable, and in few cases are there complete published accounts of their origins. Therefore a great deal has to be based on an examination of the maps themselves, rather than descriptive sources. Even so, the maps produced by particular agencies are often quite recognizable in terms of style.

In one of the few comparative examinations of topographic map series, Piket (1972) draws attention to the fact that even in those cases where the landscape is essentially similar (as in much of northern Europe), both the content and representational methods of map series produced in different countries show interesting variations. Therefore the interpretation of the topography, the degree of detail in classification, and the emphasis given to particular aspects reveal considerable differences. This is also influenced by the surveying methods used. If map series at the scale of 1 : 50 000 in several European countries, and their counterparts in the United States (including the 1 : 62 500 scale) and Canada are taken as examples, there are distinct national differences in style. Much of this may be dependent on the characteristics of the topography itself, as there is obviously a visual contrast between maps of alpine regions and maps of the North European Plain. Even so, as Piket observes, changes in landscape type do not coincide with international boundaries, and the same landscape is represented differently in maps produced in adjacent countries.

Landscape characteristics

The general visual appearance of any sheet in a topographic map series is a function of both type of landscape and its graphic representation. If the landscape is dominated by the presence of extensive features, such as water, forest, mountains, or urban areas, then these are bound to exert a strong effect. For example, some Canadian sheets include little more than water, contours and forest. Some landscapes are relatively monotonous; others offer a great variety of features, and the character of the landscape may reside in their interrelations. There are considerable differences in the degree to which particular topographic features are classified and symbolized, and thus the way in which the character of the landscape is expressed. The size of the country concerned is an important factor. Large countries, such as the United States, have a far greater variety of geographic and topographic characteristics, the nature of which must be accommodated by a standard specification.

Virtually all medium-scale topographic map series are based on a 'classical' style which evolved with the introduction of lithographic printing. This made possible the use of different hues to represent and contrast the major feature categories. In its simple form this is based on black planimetry, blue water, brown contours and green vegetation. The early 1950s Swedish series used only these four colours. This style has been expanded and elaborated by the introduction of more colour contrast, initially by the addition of red for emphasis. Most topographic map series in this scale group are therefore variations on this classical model. This is affected by the detailed treatment of particular features, especially relief, roads, built-up areas and land type. The map design is a reflection of content as well as graphic expression, and it is impossible to separate the two.

Some comparisons between a selection of European and North American map series at this scale will illustrate the point. The treatment of relief varies from the use of contours and heights, as in the British, Swedish and standard United States series, to the reinforcement of the three-dimensional impression by the addition of shading, as in the French map series. The most sophisticated and elaborate representation is the Swiss, using contours in three colours (black, blue and brown), detailed rock drawing and hill shading. Road classifications range from eight on the British series to four on the United States specification. Some, like the Danish and Swiss series, include considerable detail of road widths, whereas the United States series gives no information on this aspect. The continuous built-up area is shown by a single colour tint on both the United States and British maps, whereas on the latest (experimental) Swedish design there are four distinct categories, reflecting building density, building height, and predominant use. The Danish design distinguishes between the central urban area, which is usually continuously built-up, from the built-up area in general, which may contain considerable variations in building density and height.

Probably the biggest contrast lies in the characterization of the land surface, including vegetation and agricultural land use, as these elements are represented by area symbols which can strongly affect the overall appearance. Although all these series include forest (in green) and semi-permanent planted trees such as orchards, the degree of detail indicates not only the local variations but also differences in approach to the composition of the map. For example, the French series includes not only woods but brushwood, plantations, vines and rice; the United States series includes scrub, mangrove, wooded marsh and vineyards as well as woodland and orchards. The Dutch series distinguishes between woods, brushwood, meadows with ditches, plant nurseries, orchards, arable land, heath and sand. The Swiss map adopts a different approach, the areas covered by soil (implying either natural vegetation or cultivation) being represented by brown contour lines.

The treatment of other land types shows very large contrasts. The British map gives minimal information about low-growing plants, either natural or planted, resulting in a great deal of uncharacterized white 'space'. The Swedish map includes marsh, swamp (impossible to traverse on foot), marsh with trees and forested marsh. Cut-over forest is distinguished from other forested areas. Similarly, the modern Danish series shows heath, marsh, reeds, swamp and bog.

A map covering the whole of a subcontinental area, such as the United States 1 : 62 500 fifteen-minute quadrangle series, encounters a far wider range of geographical types and environments. It also tends to reflect the nature of settlement and development. Although it follows the 'classical' model in terms of general design – black planimetry, blue water, brown contours, green vegetation, with red for emphasis – it needs to include a considerable range of specific details. Given the much lower density of settlement over large areas, landmark features become more significant. For example, there are point symbols for such things as mine shafts, water tanks, and gas and oil wells; area patterns for tailings, distorted surfaces and gravel beaches; and area symbols for submerged marsh, mangrove, wooded marsh and scrub, as well as woodland and orchards. Symbols for water features have to include intermittent streams and dry lake beds, as well as inundated land. The distinctive treatment of township, range and section lines, with red line symbols, reflects the history of settlement, for which there is no corresponding element in the European maps.

These examples should make clear that far from being a simple description of 'what is there', the medium-scale topographic map is based on a judgement of what is significant in the landscape. The marked contrasts between the different series, even between adjacent countries, demonstrate that there are no simple objective criteria for determining the content of a topographic map. But once the surveying specifications and the subsequent generalizations for derived maps have been established, it is difficult to change them without undertaking a completely new basic survey, and therefore they generate their own inertia.

Colour

Both the choice of particular colours and the ways in which they are applied in symbols have a strong effect on the overall visual impression given by these different map series at or about 1 : 50 000 scale. In all cases, the cultural information, including buildings, roads, boundaries, and names, take precedence over the natural features, but the degree of contrast between the two sets of information ranges from strong to slight. This in turn determines whether or not the map appears as a continuous surface, or as a collection of separate features. It is particularly affected by the choice of individual hues, especially red, brown, blue and green.

The treatment of roads and built-up areas, which often have a pronounced effect on the overall composition, is a case in point. The United States and Norwegian sheets use a full red for the major road colour infill, the Canadian, red and orange. The French sheets have main roads infilled in orange, with the secondary routes in an orange pink. The Dutch sheets use a light red and yellow ochre. The Danes use pink for major roads, the British red (apart from motorways in cyan), but in addition the British map distinguishes two categories of minor roads by orange and yellow infill respectively, and is the most elaborate map in terms of road classification colouring. Both the early (1950s) Swedish and the Swiss maps leave all roads white, depending entirely on variations in road symbol form and dimensions for different categories. In the Swiss map, which is both shaded and land coloured, the roads are masked out, leaving a very slight contrast with adjacent areas. Although the earlier Swedish design employs relatively heavy black lines for the road symbols, they still lie visually on the same plane as the rest of the information.

It is interesting to note that the British design tries to create a new convention by connecting the motorway colour (cyan) to the use of blue for motorway road signs. This association of ideas is defeated by the fact that the motorway symbol on the map should lie visually at the highest contrast level in the road hierarchy, which for the other roads is expressed by the relative contrast value in descending order of red, orange and yellow. Although cyan may have a logically proper iconic association with the colour used for motorway signs in Britain, it is graphically inappropriate in the composition. Even a light purple (cyan plus magenta) would have retained some visual connection between the motorways and other roads.

In general, the British map tends to make full use of colour contrast to distinguish line symbols, for example by employing four different colours in road classifications. Conversely, the newest Swedish design tends to use colour contrast more in relation to area symbols.

For the built-up areas, the early Swedish design had black outlines with black line (grey) shading, distinguishing between residential and the industrial/commercial areas. This general approach has been further developed in the latest experimental version, with magenta for the completely built-up areas of a city or town; medium magenta for areas characterized by tall buildings; light magenta for areas of low buildings (mainly residential); and light grey for industrial areas, with individual buildings in black. It is the most elaborate classification of urban areas in any series at this scale. In a sense it tries to suggest something of the three-dimensional character of the characteristic built-up area.

The French series uses two greys (black tints) within black outlines. Outside the high-density built-up areas, buildings are much less generalized by combination than, for example, on the British sheets. The Danish sheets also have grey (black tint) urban areas with black outlines, but the major urban town centres are reinforced with pink over the grey.

In contrast, the British map uses a very pale orange – so pale that it looks slightly brown in natural light, though slightly pink in artificial light. The outlines are in black. Although the standard United States series shows continuously built-up areas in pink without any bounding line, it seems possible that this may be replaced by grey, if it follows the changes introduced on other series. The Dutch map has strong pink (red tint) built-up areas with red buildings. The effect of this is striking. It only works at all because the smaller proportion of black and grey in general, and the presence of light green, dark green and yellow ochre over large areas, reduces the potential contrast of the red with white. The full use of area colour also means that white represents arable land, thereby making use of white as part of the composition.

Typographic style

The choice of type for lettering on these map series reflects an interesting change in fashion. All the older designs (Dutch, Swedish, United States, Canadian) make considerable use of serif typefaces; some of the newer ones (British, Danish and later Swedish) hardly any. In the Swiss and Dutch series lettering is predominantly in serif form. What is more unusual is that the type on the Dutch map is in the 'modern' variety of serif, with a marked variation in line thickness and a vertical stress. Even the shaded outline variation is introduced. The Canadian map uses serif styles (both roman and italic) for settlements, places and physical features; only administrative district names are in sans-serif. Whereas on the earlier Swedish map all settlement place names were in serif style, with water and land physical features in sans-serif, the new Swedish design has also followed the general trend to sans-serif lettering throughout. The Danish and British maps are almost entirely sans-serif, the only general exception being the lettering used for ancient monuments on the latter. Place names on the Danish sheets make greater use of condensed versions, and a higher contrast in weight as well as size between names of large and small places. The overall effect is that the lettering image is placed at a higher level visually and is more sharply separated from the background.

On the British map (and the modern Swedish version) it so happens that the stroke thickness of the sans-serif lettering of the small names is very similar to the line thickness of built-up area outlines and roads, which does little to separate the names from the surrounding information. One of the advantages of the Dutch map is that in minimizing the use of black and grey in the continuously built-up area, the names in black have a high level of contrast.

This marked transition from serif to sans-serif lettering styles is an interesting topic. It has been affected by the changeover to the use of typeset lettering, whether mechanical, photographic or electronic, and the appearance of many good new designs of sans-serif faces. Yet the greater range of line weight variations in serif lettering seems to fit into map designs that make considerable use of small variations in gauge and form for line symbols, like the Swiss. Could it be that the serif faces are too decorative, or too 'artistic' for an age demanding hard information and plain speaking? They are no less legible, and as the Dutch map shows, fine serifs are not beyond the reproductive capacity of modern lithographic printing, even at small sizes. Or could it be that a 'modern' design has to find some way of being different? The change to sans-serif styles does not alter the information, and having less variety, is more monotonous. It is doubtful if the average map user is aware of the characteristics of different typefaces. Yet somewhere there has been a compulsion to move away from what had been done before.

Map and landscape

These topographic map series at 1 : 50 000 and 1 : 62 500 scales show considerable contrast in the use of colour, treatment of landscape features, and overall appearance. At one extreme, the early Swedish map, with its cool colours (black, blue, brown and green), had a sombre air; it may well have been in tune with a Northern landscape, especially one which in large part is characterized by rock, forest and water. Yet apart from hill-shading, and the different blue and green, it is very like the previous French (type 1922) series, which itself represented a long European tradition. At the other extreme, the Dutch map series is vividly flamboyant. In a country with such a high density of population and development, where all land is precious and its use carefully controlled, the light red and yellow of the roads, the red of the buildings and the pink of the built-up areas stand out against a background in which the light green of the meadows and the blue of the ever-present dykes and canals are dominant. Perhaps only a cartographic tradition as long and eventful as the Dutch can provide the confidence which such a design requires. It is in marked contrast to the standard Norwegian, Canadian and United States maps, which are closer to the 'classical' tradition. Perhaps because the style is so familiar, they are less exciting.

Three of these European maps deserve particular attention, two new, one old. The modern Danish series is altogether quieter in style than the Dutch, but manages to retain a good level of contrast within a generally subdued design. The choice of pale red, green and blue are especially important. It is a very detailed map, and like the Dutch, cannot make use of the representation of relief to make it more dramatic in character. The detail of topographic features, the balance of line gauges and forms, and the variety in weight of the sans-serif lettering give it a modern but elegant appearance.

The experimental (fifth) edition of the Swedish map will be fully digital in production. Unusually, five variations of the colouring for the urban areas, together with a complete description of feature classifications and their map symbols, were issued to a number of users and cartographers for critical comment. This followed an earlier investigation which was concerned with such questions as whether forest should be shown in green, or left white. In a heavily forested country the use of a continuous light green symbol can be regarded as less informative than in regions where more limited forest areas are visually identifiable on the map sheet. This is also influenced by the familiarity of the standard orienteering map specification in Sweden, in which forested areas are white. Significantly, respondents were almost equally divided in their opinions, which demonstrates that consulting prospective users does not necessarily solve design problems.

Of all these map series, the Swiss is the one which comes nearest to giving the impression of a continuous topographic surface. The individual symbolized features, such as roads and buildings, exist on the land surface, not on a different plane. This is achieved by using shading and land colour in combination. Even the serif lettering seems to be part of the landscape rather than an external addition. Although the style reaches its climax in the sheets covering the Alps, it is just as effective in hilly and level country. In general, symbol dimensions are relatively small, so that the detail needs attention. The land background, predominantly yellow–grey–green outside the Alpine ridges and glaciers, provides a sufficient colour contrast with the cultural information, which is almost entirely in black. With contours in three colours, detailed rock drawing and broad hill shading, it is graphically sophisticated and technically elaborate. It is the type of complex and subtle design that has to be absolutely right, or it would be disastrous. For example, if the shading was slightly stronger, or slightly weaker, it would be quite unsatisfactory. Cartographically, it lies at the end of a long European tradition, rather than at the beginning of a new one. It both appeals to, but make demands on, the map user. Not surprisingly, its primary originator was also a great landscape painter, and its expressiveness lies in its power to evoke the landscape. There is a correspondence between the subject matter and the representation, a 'satisfying appropriateness', which is rarely encountered. Like all artistic works, it arouses both admiration and criticism.

Style and general atlases

The general world atlas, produced mainly for educational purposes, is also a cartographic product that becomes familiar to large numbers of people through its use in schools and colleges. Indeed, it may contain the most familiar of all maps to many members of the general public. Although devised primarily for learning, virtually all such atlases also serve as general reference atlases, or provide much of the basic material for such atlases sold through the book trade. It is interesting to compare three typical examples of this kind of atlas.

The three atlases are the *Atlas Advanced*, published by Collins–Longman (1968); *De Grote Bosatlas* published by Walters–Noordhoff in the Netherlands (1971); and the *Atlas Scolaire Suisse* published by the educational departments of the Swiss cantons (1972). In the Collins–Longman atlas the maps in some sections, including those of the Americas, are produced under copyright agreement with the Rand McNally Company of the United States.

Like all such atlases, there is general coverage of continents and countries, together with a large number of specialized maps on particular topics. Each atlas has a recognizable style for its general maps, which is maintained throughout the work.

Each atlas contains a map of the United States at a similar scale, ranging between 1 : 12 000 000 and 1 :15 000 000. Relief is represented by a combination of hypsometric and bathymetric colouring, with hill shading, together with the usual cultural information about boundaries, transport, towns and cities. The three atlases differ considerably in appearance, and in the relative emphasis given to the major elements in the information.

The Dutch atlas is bold, with layer colours running from dark green at the lowest land level, through two other greens, green-brown and three darker browns, making a total of seven zones on the land area. The colours are comparatively dark, and do not follow any graphic principle, as they are relatively darker on both lowlands and highlands than in the middle of the range. Even so, there is quite a small lightness variation, and the contrast is achieved mainly by differences in hue. Inter-layer contrasts are small, in order to make a smooth progression from one layer to the next.

The Rand McNally map is altogether lighter in the treatment of relief, the land layer colours following a logical progression from light green through yellow-green, light brown, yellow-brown, yellow and white.

The whole effect is noticeably lighter than in the Dutch map. The Swiss map is noticeably 'cooler'. The layer colours run from a dull green through light green, pale brown, yellow, light yellow and white (the higher, the lighter).

All three maps use grey hill shading, but with a different effect. In the American map, the shading is relatively light, and is limited to major mountain ranges, with no attempt to model the relief in the less steep areas. As there is no overall land tone it is not possible to bring out the light-facing slopes, and so the relief effect is limited. Both the Dutch and the Swiss maps use a full oblique shading, adding grey tone to the level areas, and thereby accentuating the light-facing slopes. Because of colour choice for the layers, the Dutch map produces a rather strong image, relieved partly by the yellow, and the use of a reddish brown for the highest elevations. In comparison, the Swiss map is much cooler, and the major mountain ranges stand out boldly. Whereas both of these maps omit contour lines, limiting the hypsometric layers on land, the American map includes them.

The combination of hypsometric colouring and hill shading of course poses a classic cartographic problem, to which there is no perfect answer. On the American map it is possible to identify the individual layers, but the shading effect is deliberately limited. On both the Dutch and Swiss versions, the shading is bold, but it becomes virtually impossible to identify the layers in strongly shaded areas. The American design gives more discrete information about elevation; the other two sacrifice this for a smoother progression between layer colours. The European designs attack the problem of combining a complete shading with the colours; the American design tends to avoid the problem by reducing the extent and intensity of the shading. These are, in fact, incompatible requirements, and no design can satisfactorily encompass both.

The second element in the overall design is the treatment of the cultural information. In the American map this is very much in the foreground. The red state boundaries, red built-up areas and red railways lie visually above the land surface, and are reinforced by using bold condensed sans-serif capitals for state names. The Dutch design again uses a high contrast level, with heavy black railway symbols, fine red lines for roads, and red infill for town symbols. The Swiss map is quite different. Boundaries are in green, railways are in red, and both roads and town symbols are in brown. Both the Swiss and Dutch maps use lower case letters for state names, which are visually much less prominent. This treatment of the cultural information is reinforced by the selection and density of names. For example, in California the American map includes over twice as many town names as either the Dutch or the Swiss.

The styles of the three designs are clearly influenced by both cartographic and non-cartographic factors. For the European maps, the cultural detail of the United States is less important at this scale, and therefore is given a much more selective treatment. At the same time, the importance of physical geography in teaching in European schools and colleges almost certainly affects the emphasis of the map as a whole. Both the European maps are predominantly concerned with physical features. The American is a general map in which state and city names are important: the relief is only a background. Both the European maps are influenced by the cartographic styles common in their respective countries. In its treatment of relief, the Swiss map follows the principles of the 'Swiss manner' established in its topographic maps. The choice and balance of colours is very carefully thought out and is graphically consistent. The use of green for state boundaries is certainly not a common convention, but it is visually correct in this particular design. In this respect, the cartographic considerations are dominant. The Dutch map has the strong colouring characteristic of so many Dutch products. In the American version, the pale colours used for elevation introduce little interference with the cultural information. It is quite possible that this level of relief representation goes as far as is deemed sufficient, as it is designed primarily for the American market. All three maps are legible. All three in their own way provide a norm against which other small-scale atlas maps may be judged; and all three are quite different in style.

Factors in cartographic style

There are many reasons why the map products of particular organizations should demonstrate individual styles. Some of these must lie with the cartographic interests and ideas of individual cartographers and map publishers, whether or not they can be associated with particular people. It is clear that many individual cartographers have distinct mannerisms in that they have a preference for certain colours, typographic styles, treatment of borders and so on. To some extent these tend to be perpetuated by experience. Having found that a certain graphic element or composition is successful, it is subconsciously adopted when a similar task arises on another occasion. Unlike paintings, maps are strongly influenced by technical and production factors, many of which have changed with the adoption of different technical processes and working methods.

Production methods

It has been mentioned already that engraving had a strong influence on the appearance and design of maps produced over a long period. As the entire image, including lettering, had to be cut into the copper plate, the difficulty of doing this to a high standard inevitably meant that the required skill could only be achieved by long and sustained practice, and even then the construction of a large map needed a great deal of time. Engravers therefore developed a style of working which both exploited the medium, essentially the ability to produce fine, sharp lines, and operated within its limitations. Therefore the types of lettering that could be produced successfully by engraving had their own characteristics. The need in particular to engrave small names to a typographic and size standard inevitably meant that the engravers developed a particular style, and this in turn was taught to apprentices. Thus the similar appearance of so many engraved maps is essentially a function of the difficulty of the medium.

Once the requirement for much larger numbers of maps developed, then the engraving process was both too slow in production of the image, and inadequate for the rapid printing of many copies, because of wear on the plate. Despite the introduction of many different kinds of aid – etching, dye stamping, electrotyping, etc. – lithography offered obvious advantages. Initially there was considerable difficulty with the quality of the line image produced by early lithographic plates and presses, and indeed lithography was often disparaged in comparison with engraving in its early days. As lithography improved, so the range of graphic possibilities increased.

With the need to produce large numbers of sheets in a map series to a common standard, it was necessary to ensure that all the cartographic draftsmen could produce standardized symbols as required. Therefore there was an emphasis on this in training. The particular style of the organization was the outcome of this conventional approach; conventional because the adoption of certain types of symbol and certain type of lettering, successively taught to its employees, provided its own inertia. The style sheet was laboriously copied many times by the apprentice draftsman, and the need to maintain this was ingrained. It was particularly true with regard to lettering, which was always difficult to standardize.

In addition, there was a premium on those methods that could lead to the most rapid production. The double-line road casing, now so commonly used, was extremely difficult to draw in ink before the introduction of double-line pens and working at a larger scale for subsequent photographic reduction. The advent of double-line scribers eased this problem, and a range of double-line road symbols is now widespread. Even so, the difficulty of producing correct junctions for cased roads on automatic plotting machines is having an effect, and several of the newer fully digital map series are abandoning this type of line symbol.

Technical innovations and changes have a profound effect on map style, mainly by virtue of what they offer in the sense of graphic possibilities. The light, clean appearance of so many modern maps is a consequence of the great improvement in lithographic printing – presses, inks, plates – during the last thirty years. They offer a sharp comparison with the heavy, dull apearance of many maps printed in the 1930s. The acquisition of large-format percentage screen tints revolutionized the production of area colour, and made a refined treatment of both individual tints and colour combinations possible. Photoset lettering based on a full range of standard typefaces had a similar effect on the quality and range of map lettering.

On the other hand, technical facility can introduce its own dangers. The ability to produce fine lines by scribing, maintained by high-quality reproduction, can lead to an excessive and unneccessary concentration on fine lines, simply because they are technically possible. What is now technically possible frequently outruns what is desirable in a graphic sense. To some extent this may be a consequence of the unspoken assumption that 'fine' equals 'more accurate' or 'higher quality'. Illegibly small lettering, excellently reproduced in print, is a trap for the unwary.

The factor which inevitably introduces a degree of inertia in cartographic style is the nature of the investment in the original images. The national survey, with possibly many thousands of sheets to produce, establishes a particular style for the map series. As it may continue in production and revision for decades, it leaves little opportunity for any fundamental changes in design. Many of the European topographic medium-scale series, originally created during the period of engraving, continued virtually unchanged for most of a century.

The same problem affects commercial publishers. The creation of a major world atlas, for example, which is often characteristic of a commercial map publisher, involves a large investment of capital. Therefore this input is maintained and exploited in different versions over a long period. Although minor 'cosmetic' changes may be made for derived products – such as slight changes in printing colour – the basic structure of the graphic representation will not be altered in any significant way. Even though the creation of digital databases should make it much easier to produce different graphic outputs from the same coordinate data, the association of a particular company with a particular, and often familiar, style may well influence any new designs.

Style and user inertia

The more subtle consequence of the influence of particular map styles lies in the user's experience and familiarity. In many cases individual map users become familiar with certain types of map through the products of a single agency. Consequently these products tend to set a norm, which in turn leads to the expectation that other maps will be similar. It is clearly evident in the way that many map users regard the products of their national mapping agency as typical, as standards against which other styles are subconsciously judged. Being generally unaware of the existence of alternatives, what they are used to affects their expectations. Consequently, having created this expectation, it may be difficult for the producer to introduce a completely new style with any confidence.

The description of cartographic style

Given the importance of technical processes in influencing the style of printed maps, there has been surprisingly little attempt to either describe or classify stylistic differences, even when pronounced technical changes have affected the appearance of maps. The ink-drawn map in which the entire image, including lettering and patterns, was drawn by hand, existed over a long period. Although this style contained minor variations, such as the use of stamping for lettering, or the production of patterns by ruling on the plate, it was essentially manual, in the sense that the images were drawn first and reproduced subsequently in printed form. There was a simple distinction between constructing the map, and then reproducing it by making a set of photographic negatives, and using these to copy the images onto the printing plates. Thus there was a specifically 'manual' stage in the whole production sequence. Yet cartographically there is no accepted term for 'manually constructed/lithographically printed'.

By a curious anomaly, now that maps can be produced either wholly or in part by digital methods (although they are generally printed by lithography), in which there is no manual construction of images, it has become necessary to distinguish these products from the current non-digital procedures. Frequently the non-digital methods are referred to as 'manual', 'traditional' or conventional', despite the fact that these methods, which have developed over the last forty years, are a combination of manual image generation with a variety of reprographic processes, subsequently printed by lithography.

The lettering, point symbols and patterns on such maps are not individually drawn by hand, as they were in the period of manual lithographic production which preceded them. The colour tints which are so important in all kinds of modern maps could never have been produced by manual methods – a technical and graphic difference which in itself is enough to distinguish the modern period from the purely 'manual'. This hybrid style, an often complex combination of manual and reprographic methods, has no accepted name in cartography, despite the fact that its graphic sophistication not only sets the standard for what people now accept as the colour-printed map, but is also the graphic standard which digital methods have had to emulate. Indeed, an apparent unawareness of its existence must lie behind the erroneous supposition that maps are still 'drawn' and then 'reproduced', as two distinct and separate stages.

It could be that the prevalence of a given set of technical methods has tended to dominate cartographic production over long periods, and because these methods are almost universally applied, they are simply taken for granted. Only when a technical 'revolution' takes place does it become necessary to distinguish one type of production from another. It may also be the case that as digital map production so far has not established any significant difference in graphic style from manual/reprographic/lithographic methods, it is only the technology which has to be distinguished by a separate description.

It is possible that lithography, with its wider graphic possibilities, does not exert the same dominating control over the style of maps as did engraving, and so is not an element which itself needs to be taken into account. It is also possible that the description of maps by scale, area and subject is sufficiently specific and relatively easy, compared with the difficulty of comparing and describing individual paintings or other products of fine art. Even so, the stylistic differences between maps of the same class are worthy of emphasis.

Given the absence of accepted cartographic terms, the topographic map series examined earlier could be considered according to the stylistic terms familiar through painting and the fine arts. The basic colour composition of black planimetry, blue water, brown contours, green vegetation, and red for emphasis might be regarded as the 'classical' form, despite the many variations on this familiar theme. The Swedish series of the 1950s is the most severe – perhaps the equivalent of Doric. The Dutch is the most vivid and could be considered to be Romantic in approach. The Swiss, with its emphasis on landscape character, has some elements of the Impressionist. The French map, with its strong national tradition, is as obviously individual as the English water colour school of painting.

The same approach could be applied to a wide range of specialized maps. For example, the stylistic periods of nautical charts are even more evident than those of topographic maps. Although the analysis and description of style might be regarded as serving no practical purpose, it would at least make the point that the functional description of a map does not say everything about it, and its design and appearance are also worthy of attention. Perhaps the historians of cartography can devise an appropriate terminology.

PART FIVE

The map as a product of skill

CHAPTER 18

Digital maps and geographic information systems

It may seem paradoxical that as map making has entered a digital era, questions of the relationship between cartographic design and art are being looked at with increasing interest. Although many would not express it in those terms, the difficulty of controlling design in a manner compatible with the technological power of digital systems is raising old questions in a new form.

In order to understand how this position has been reached, it is necessary to consider what roles knowledge, skill, creativity and artistic ability may play in a changed environment; not just in cartographic production but especially in many other geographic information systems in which the output of maps may be a necessary or at least desirable feature.

From a broader perspective, the use of the computer to store, analyse, process and output information has become so widespread that ideas and methods originating in diverse fields are being examined to see whether they can be applied to aspects of making and using maps. This in itself is by no means unusual, because map making has always drawn on technical developments initiated in other fields. Indeed, in some ways it parallels the developments of the graphic arts revolution of the 1950s and 1960s, which profited from the remarkable progress of graphic arts technology. In other respects, the questions being asked of the role of the cartographer return to some of the theorizing about cartography and map use of the 1970s and 1980s.

Whatever their origins, they all encounter one central question: to what extent can digital systems imitate or even surpass the intellectual and aesthetic functions of the human brain, and how far are such developments proper or even desirable? Given such challenges, there are naturally great differences in opinion. But the fact that such questions are encountered also makes clear that both map making and map use reflect a wider social, technical and cultural background.

The comparison between the computer and the human brain continues to be argued and frequently disputed. It compares the human properties of intelligence, perceptiveness, skill, knowledge, creativity,

individual differences and inconsistency with the computer characteristics of computing speed and power, consistency, uniformity, dependence on one type of input (digital data), and lack of intelligence. On the
one hand, the computer seems capable of tasks which far exceed the
potential of the human being; on the other hand, it finds apparently
straightforward and mundane problems extremely difficult or
impossible.

Given that so much of the normal field of map production can be
dealt with effectively by computer processing, it would seem that the
remaining area of composition and design should also be amenable to
computer control. If the material content of any map can be stored,
selectively manipulated and reordered by computer processing, and if
computer plotting and drafting can replace any manual drawing, what
then is left? Discussion of these possibilities inevitably leads back to the
role of art in cartography.

In order to put these questions in perspective, it is necessary first of
all to review, at least in outline, some of the major changes brought
about by the digital transformation, and the ways it is affecting both
map making and map use.

Digital maps and geographic information systems

Although nowadays all digital map production systems tend to be
subsumed under the general title of Geographic Information Systems
(GIS), it is important to recognize that this is a relatively recent
occurrence, and that this broad term covers a wide range of different
applications. Indeed, it was the long and often painful development of
digital mapping that made possible the geographic information systems
as they now exist.

Origins of digital map systems

The rapid development of computers after the Second World War made
a great impact on all those operations involving the handling of large
quantities of numerical data, and it was in this area that most efforts
were made to introduce the computer as a powerful new tool. It was
not long before it was realized that the possibility existed of using this
technology to control the operation of machines, in the first place in the
control of precision machine tools. Where the information existed in
numerical form, in measurements or statistics, the transition to digital
processing was relatively straightforward. But where the source material
was graphic, then the initial problem was complicated by the fact that
the source information had to be converted into a machine-readable
form.

In the late 1950s the major national surveys as well as many
commercial mapping organizations were preoccupied with problems of
renewing and updating old material, and replacing it with new and
derived map series and atlases. Many countries with large territories,

even those which were technically advanced, still had enormous programmes of basic mapping to complete. The advent of photogrammetry and improved surveying instruments had vastly increased the speed and output of surveying and basic map information. The bottleneck lay in the slow cartographic production processes, which still depended on manual construction for the key elements of the map image, even though they were improved and assisted by developments in reprographic technology. The desire to replace these slow drafting processes, which depended on a high level of manual skill, by more rapid means of production, led to experiments in many fields, for example by the introduction of photomaps of many types. But despite the advent of scribing, masking, photoset lettering, colour proofing, large-format tint screens, and so on, there was still a limit to the speed of production that could be achieved if appropriate standards were to be maintained.

Central to this problem were the essential characteristics of sheet maps: large format, intricate detail and the use of many printing colours. In order to convert existing material to digital format, methods had to be found to digitize the existing map information so that it could be subsequently output as finished maps, without any loss of information quality. In particular, the need to digitize the line and point images which provided most of the basic map information made huge demands on the recording rate, storage and management of the digital data.

Initially the major problem was the often experimental construction of hardware, both digitizing tables and large-format automatic plotting machines. Few organizations could support the required scale of investment and development, and the military mapping agencies, with their need for rapid response output, were an important contributor in these early stages.

At the other end of the scale, researchers using relatively small computers were already employing them extensively for statistical processing. The need initially to output tables and simple diagrams then led to the production of small-format black-and-white maps. This was especially important for geographers and other environmental scientists, who, as map authors, were also dependent on the manual drawing output of cartographic technicians, usually working with very limited technical resources. A major difficulty was that very often the potential value of a given map could not be considered until after it had been drawn, so that on many occasions maps had to be made which eventually offered little to the progress of the enquiry. What was so often needed was some means of rapid output, even if the map itself was cartographically at a very basic level.

Where a map in colour was required in order to present the information clearly, the cost of production in print was prohibitive unless a sufficiently large number was produced at one time. Given the importance of map illustrations in many types of research appearing in scientific and other academic journals, this cost factor inhibited the production of maps, even in cases where they were clearly desirable.

What was needed was some means of small-format, high-quality multicolour production that would be economic in small quantities.

In many respects the achievement of fully digital production was easiest with large-scale urban and cadastral surveys, especially after the introduction of surveying instruments and photogrammetric plotting machines which could output their measurements in digital format directly.

The technical problems were gradually overcome as computers, and digitizing and plotting machines were developed. The phase in which hardware development was paramount was gradually replaced by an emphasis on software development to improve data management and processing. But given the nature of the large-format, multicolour sheet map, with its standards of accuracy, the initial technical task was to be able to produce an image of equivalent quality, an objective which has only recently been realized.

From digital maps to information systems

As is often the case with major changes in technology, it is only during the course of technical development and experience that new possibilities are realized. Whereas the early mapping systems had to concentrate on the task of simply achieving map production, the transition to the use of digital data gradually made other possibilities evident.

It was clear from the outset that cartographic requirements that could be mathematically defined, such as the plotting of projections, could be done much more efficiently by digital processes than by any form of manual construction. For national surveys in particular, although the digitizing of separate map sheets created difficult problems in matching the data at the edges of adjacent sheets or digitized sections, once the digital database had been created, it then became possible to consider many new possibilities.

One of the long-standing complaints of many of the official users of national surveys was that the standard specification for the map did not suit local or specialized requirements, and frequently that published sheet lines were inappropriate. From a properly managed digital database, the user could output map sheets as locally required, and indeed could plot only a certain selection of the total data. If the organization also had other geographical data which were based on the same coordinate system as the source maps, even though these were normally in the form of a standard base map, then combinations of data could be made for specific purposes. It was this flexibility which opened the door to extensive geographic information systems which could be purpose built.

The impetus for this came from the users, in particular those users who had long depended on the slow and often inaccurate manual plotting of different sets of information onto a base map, especially as the base itself was frequently out of date. The need to compare the distribution of one phenomenon with another over a given geographical area was central to many kinds of study, yet it was often painfully slow and time-consuming.

Administration, management and land information systems

Although the advantages of digital systems for geographical research figure largely in the literature, it should be appreciated that the main users of such systems are those concerned with management and administration, rather than research. In Europe and North America in particular, many city and local authorities, as well as public utilities, are heavily involved in collecting and using geographic information for management purposes. Land registers, tax registers, building registers for insurance and fire risk, population data for school and police authorities, physical planning and development control – all these need current information about geographic location. Operating as entirely separate departments, there is great duplication of effort and frequently a great waste of resources. Yet in many cases, it is necessary to study many disparate sets of information to solve particular problems, or make appropriate decisions.

A number of city authorities, notably those which already had their own surveying and mapping organizations, realized that an up-to-date large-scale topographic map was the essential base on which to coordinate other sets of geographical information. As long as the base map contained the necessary ground information in terms of buildings, cadastral boundaries, etc., selected parts of this in digital format could be output as required for other users, provided that they in turn collected their statistical data on the same coordinate basis. For a long period these were identified as land information systems, because the basic element was a large-scale survey of the land surface of the administrative area. In those cities or administrative areas that had a full cadastre, the topographic base was normally combined with the cadastral detail, a key element in planning and development.

Similar possibilities also existed at national level for the coordination of statistical data based on geographic location for such things as taxation, population movements, planning and development (see Rystedt, 1977; Wastesson, 1977). The same problems arose for many individual agencies concerned with agriculture, forestry, water resources, etc., and indeed many of the most spectacular achievements came about in these applications. For most of these, however, the other factor that was to play a crucial role was the advent of remote sensing.

Remote sensing

Although the use of aerial photography had long been established for the production of plans and topographic maps, making use of the important characteristic of stereo imagery, the existence of such images for large areas naturally led to the desire to interpret the imagery itself, rather than to wait for it to be eventually transformed into maps. This was particularly the case where information was needed which did not necessarily appear on the maps themselves. Although the direct use of the photographic imagery led to many experiments in which some elements of the aerial photographs were incorporated directly into maps, for many scientists and other researchers, such as archaeologists, the aerial photographs themselves were important sources of information. Initially all the interpretation was carried out by visual inspection, involving visual search, identification and classification.

Other types of remote sensing opened up wider possibilities. Whereas aerial photography for photogrammetric purposes taken at relatively low altitudes remains unrivalled for detail in the visible spectrum, satellite-based sensing systems can measure reflected and emitted radiation at a much greater range of wavelengths. Although virtually all this imagery is in effect small scale, high-altitude airborne sensing systems have improved both resolution and the rather limited stereo effect.

Whereas the aerial photograph is a continuous-tone image (which can of course be scanned into digital format), most of the newer systems record in digital mode. Digital image processing can be used to extract particular hues or particular densities (tonal values) from the images, largely replacing, or at least assisting, the process of visual interpretation.

Given the speed and frequency of production of many of such types of imagery, it is hardly surprising that the combination of map base, remotely sensed images and other geographical data – all in digital format and all converted to the same coordinate base – provides a powerful tool for environmental research and monitoring.

What is significant is that different types of system that were initially conceived and operated quite independently gradually came to be perceived as variations on a basic type. Even so, it must be understood that the collective term, geographic information systems, covers a wide range of activities, directed to many different purposes. One of the critical differences between them is the resolution of the original information.

Resolution

All geographic information systems are based on the fact that the data concerned can be located in relation to the Earth's surface. This is the common or unifying element between them. One of the major differences is the degree of resolution of geographical position. For example, a city land information system will need to identify clearly every building, street line, frontage, administrative and other boundary, and in the case of public utilities, every pipe or cable. This can only be achieved by a survey of a suitably large scale, through which position on the land surface can be measured within centimetres.

For other puposes, this positional detail may not be needed. For example, census data may only be required for presentation on the basis of relatively large administrative areas. For such purposes, an outline of the boundaries of these administrative areas, together with major topographic features, is quite sufficient and appropriate to the requirement. This may be at a much smaller scale, and the digitizing of such boundaries is a relatively small task compared with the production of a large-scale survey. In this sense the detail of the base map input needed to define the geographical position of the data varies widely. Although on occasions the location of the boundaries may be changed in some administrative reorganization, carrying out such changes is a minor task compared with the maintenance of a large-scale base map of an urban area which will have to record changes and developments on almost a daily basis.

One of the problems of digital data is that often their characteristics are not necessarily apparent to the user. Where the base map information is stored in digital form, its true scale is invisible. So that although some writers state that such data can be output at any scale, Aronoff (1989: 139) is quick to point out that although the data may not have a specified scale, 'they were produced with levels of accuracy and resolution that make it appropriate to use them at only certain scales.' Although the problem of generalization is not referred to here, it is a pitfall for many geographic information systems which may have quite incompatible sets of data, or from which data may be output in maps at inappropriate scales.

Geographic information systems and map output

Maps play a dual role in virtually all geographic information systems. On the one hand, maps are a source of data input, and on the other hand, they are a means of presenting the results of interpretation and analysis. Although theoretically it would be possible to identify a location on the Earth's surface by simply showing a section of graticule or grid, in practice the subject matter is always located on some representation of the Earth's surface. Even though a processed remotely sensed image may serve as such a basis, normally it is amplified by the addition of key topographic features, boundaries, urban areas, and names. The ability to select such information as required from the basic digital data is an important aspect of the fully developed system.

So far as normal map production is concerned, the question of map design does not necessarily impose any entirely new problems, or even offer any entirely new solutions. In a cartographic organization the presence of experienced cartographic staff provides a continuity of knowledge and skill. Even though the particular conditions of working with digital databases rather than graphic ones, and the change in working practice, may cause a considerable upheaval, the organization will already have the cartographic expertise to deal with the design of new products. In some cases, especially for major topographic map series, the map specification is unlikely to change for long periods, because continuity of symbolism and map style is important to the user.

For organizations working for a variety of clients, and therefore constantly generating new maps, there can be many advantages in digital production, assuming of course that the digital base and operating procedures are adequate for the purpose. Chief among these is the ability to experiment with composition and design. Different scales can be tried, as well as different shapes for maps which have to be fitted into text or accompany other illustrations; and different designs can be displayed and assessed. True, if the map is to be printed by lithography, the difference between the screen image based on transmission and the print viewed by reflection is considerable. But cartographers will become accustomed to adjusting for such differences.

Much depends on the nature of the work required. Using available digital data sets can mean that the slow and often lengthy process of compiling maps by hand, especially those which involved several different sources, can be avoided. But if the digitizing process has to be undertaken to obtain the map information – or part of it – in digital form, then the cost of this may be a major factor. For this reason alone it is likely that many small and local mapping requirements may be dealt with manually for a long time to come.

The longer-term effect on symbol design itself is more difficult to assess. It is clear that some kinds of map symbol are more difficult and time-consuming for digital production systems than manual ones. For example, the demise of the interrupted line, which has already happened in some map series, diminishes the range of graphic variables for the line image. On the other hand, the development of digital elevation models and digital terrain models, despite the cost of producing them for large areas, may well make it easier to introduce hill-shaded representations of terrain than it was when every one had to be drawn by hand and subsequently converted into half-tone. Given the value of the hill-shaded basis as a background for the display of specialized information shown by line and point symbols, this change may well increase the possibility of making use of this method. Given that all technology seeks to maximize advantages in speed of output, technical changes will continue to influence symbol design, just as they have in the past. After all, there was a time when scribing was criticized because it was no longer possible to produce a tapered line, a feature characteristic of pen-and-ink drawing.

Digital mapping and map users

Given the impressive speed of development of geographic information systems and their dramatic effect on many types of research and management activities, it is easy to forget that maps and atlases are important to many people in what may be called the general public, as well as in education. Although, in due course, the potential (not yet realized) for speed of revision and updating may actually improve the qualitative value of the maps and atlases sold to the public, for many ordinary users the digital revolution is invisible. Despite the claims of the publishers, the types of world atlas and road atlas, as well as the more popular scales of the national survey, normally available to the public through book shops, stationers and specialized map outlets, remain essentially the same in character and appearance as they were in the previous production era. This is hardly surprising, given that the mysteries of production are of no interest to such users, unless they affect the visual characteristics of the maps and therefore their familiarity, or their published price.

Now it is true that most map-making organizations using digital systems are in the phase of creating and developing digital databases and the operating systems required. For many, such investment is a heavy burden, and it is likely to be a long time before any advantages of more efficient production can be converted into lower prices. This might eventually have an effect on the market for these types of map publication.

Where currency of information is vital, as for example in street guides for towns and cities, it is more likely that customer-controlled display systems may partly displace the printed map. Although this has been forecast many times, there are still many practical advantages in the paper copy which can be scrutinized, returned to as needed, and carried about as required. If hard copies can also be obtained from data sources which are maintained fully up-to-date, then this would be a distinct improvement in map information for such users.

Map output without cartographers

It is in this field that the most interesting possibilities – and some of the most intractable problems – exist. Broadly speaking, there are two major groups likely to be in this position: researchers using databases, and non-specialist map users. The former are primarily those carrying out research using geographic information systems; the latter are the amorphous and undefinable group of people who are familiar with personal computers and computer graphics. The so-called 'mapping package' is becoming increasingly common. For them a map is probably something between a diagram and a picture. As geographic databases become ever more widely available, it is probable that some of these users will venture into making maps.

What both groups have in common is that any map is likely to be only one of the alternatives available for graphic display. The concern, therefore, is with presentation and illustration usually allied to text, tables and diagrams. Given the present fascination with personal computers, the very fact that the possibility of making maps exists is in itself likely to entice at least some users in that direction. Whereas the first group is almost certain to be aware that there are problems in cartographic representation, the second group may be quite unaware that anything called cartography exists.

For some specialists the possibility of producing maps without cartographers was always seen to be one of the primary advantages of digital systems. This was not really due to any bias, but in effect expressed a desire to be free from dependence on slow manual output, and often the difficulty of conveying clearly enough to the cartographer what was wanted. In addition, many cartographers had an annoying tendency to talk about cartographic principles and even 'good design' – matters which many intending map authors regarded as irrelevant.

For others, the slow and often tedious business of putting map material together in graphic form, at least to the point where it could be given to the 'cartographer', was always a source of difficulty, especially for those unfamiliar with cartographic methods and practice, and unskilled in cartographic tasks.

For both these groups, digital mapping offers interesting possibilities but also a range of usually unforeseen problems.

Visualization and map use

The role of visualization in map use has long been a subject of interest to those concerned with how people learn from maps, and how map structure and design affect this interpretative process. The ability to extract, process and transform data from digital databases is one of the most powerful tools of analysis in many types of geographic information system (see Hearnshaw and Unwin, 1994). A consequence of this is that the term 'visualization' has taken on an additional meaning. As MacEachren (1991: 154) puts it 'Visualization is both a human ability to develop mental images (sometimes of relationships that have no visible form) and the process of rendering concrete visual representations designed to assist this mental visualization process,' and he refers subsequently to 'rapid growth in interactive scientific visualization tools' (MacEachren, 1991: 161). But it is important to remember that the term has two connotations.

Given the capacity to interrogate the data for a given area, changing orientation, scale and selection, summoning not only maps but views of three-dimensional features and two-dimensional sections and diagrams, not only would map interpretation be enhanced, but the rhetorical nature of a particular map as only one out of many possible choices would be made more obvious. Such a development, if it could be pursued through school education in particular, might induce a much more critical understanding, especially where the subject matter involves numerical values and statistics.

Maps, diagrams and images

The facility with which different types of graphics can be produced is also having an influence on attitudes to maps. In some respects, it is just as easy for incompetent users to create graphic disasters as it is for them to create inadequate maps. Given the fascination of the media with dramatic graphic effects (for example, all those television programmes which are preceded by impressive but totally irrelevant graphic displays), it is not surprising that even those individual users who are visually unskilled should want to make use of them. This has two major consequences. It has been pointed out by many researchers that the creators of most general graphics systems show little comprehension of the controlled and deliberate use of the graphic variables which is typical of cartographic practice. Far from cartography learning from them, they have a great deal to learn from cartography. As Fisher *et al.* (1993: 141) put it, 'many visualization systems ignore some of the simplest rules.' In a rather different context, Wood (1993: 153) observes that 'knowledge of basic design is essential to provide programmed expert support along with other black box facilities for mapping systems such as GIS.'

The other, and more subtle, consequence is that the distinctions between images, diagrams and maps are being blurred, a matter which is of more consequence to the cartographic community and to map users than seems to be realized. This has even gone so far as advocating the treatment of cartography and map making as no more than an aspect of graphics in general. The geometrical structure of the map, which is an essential characteristic no matter how remote it may seem from a simple outline of a region, is the foundation of the relationship between the map and the Earth's surface. Coordinate systems may not be shown on many maps, and may not be necessary, but all the simplified and selected map outlines at some point have been derived from the observation and measurement on which basic maps are founded. If this is ignored, then the necessary belief by the map user that what the map states can be correctly located on the Earth's surface will be diminished. Although the notion that both cartography and map need to be redefined seems to be imported into cartography by enthusiastic outsiders once every decade or so, the discipline of actual map making is essentially different to the methodology and objectives of other types of graphic communication.

All these developments focus attention on a central question: to what extent can digital systems and computer programs take over functions presently carried out by human operators? It is obvious that a whole range of both graphic and cartographic devices are already well established in both geographic and cartographic systems. Much if not all map production is carried out by computer-controlled machines. A great deal has been learned about visual perception and map interpretation, although neither is fully understood. But if programs can be written to control such a variety and range of output, is it conceivable that the whole map-making operation could be carried out by a properly developed geographic information system? This would have to encompass both composition and design. It is such a challenge that it is hardly surprising that in an age when artificial intelligence is often perceived as the next great step forward that its possibilities should also be considered for making maps.

CHAPTER 19

Artificial intelligence and cartographic practice

In order to produce a map by digital methods, whether for normal cartographic production or within a geographical information system, a sequence of operations has to be carried out. Whether the production is in analogue or digital mode these are fundamentally the same, even though the terminology is often different. The current position in nearly all cases is that part of the sequence can be carried out directly by computer-controlled operations, and part depends on decisions and actions controlled by a human operator. The interesting question, therefore, is whether eventually the digital system can be developed to the point where it would be capable of carrying out all the procedures independently, doing without any human intervention apart from the initiation of the project.

Recent and current cartographic research efforts in the application of artificial intelligence fall into two main groups. On the one hand, there has been a sustained interest in algorithms and programs that are capable of performing specific cartographic tasks, mostly related to the preparation of digital data for new maps. Among these, aspects of generalization and name arrangement feature prominently. On the other hand, attempts to produce complete maps from a digital database, with varying degrees of human assistance, fall into the category generally known as expert systems. There is a considerable overlap between the two. Both are linked to the whole concept of artificial intelligence, in which computers take over tasks that were formerly regarded as dependent on human intelligence. To understand how this may be applied to map making, it is necessary to at least review briefly some of the basic considerations that lie behind these developments, because they raise fundamental questions about what terms such as 'intelligence' and 'knowledge' really mean.

Artificial intelligence

If an artificial creation is to be regarded as intelligent, then it is necessary to consider what the term 'intelligence' represents. As is often the case, general terms in common usage are capable of sustaining many differences in meaning. The groups of definitions in the *Shorter Oxford English Dictionary* introduce a number of components or properties, referring to 'the faculty of understanding, intellect ... understanding as a quality admitting of degree ... the action or fact of mentally apprehending something; understanding, knowledge, comprehension.' In turn, intellect is described as 'That faculty, or sum of faculties, of the mind or soul by which one knows and reasons (excluding sensation, and sometimes imagination)'; and knowledge is given several definitions including 'Acquaintance with a fact; state of being aware or informed; intellectual acquaintance with, or perception of, fact or truth; the fact, state or condition of understanding.'

Despite the single reference to perception, it is clear that in normal contexts intelligence is regarded as making possible knowledge and understanding, and that it is primarily regarded as an intellectual property which is acted upon consciously. In normal usage, intelligence is essentially the capacity to understand even difficult propositions, and to bring to bear a rational mind, not hindered by ignorance, prejudice or emotional states.

If knowledge is dependent upon intelligence, then this in turn raises several questions: how is knowledge acquired, and to what extent can it be shared between people or communicated?

This, of course, is a problem which has been debated by philosophers through the centuries. Opinions range all the way from conceiving the mind as a *tabula rasa* onto which all knowledge of the external world has to be imprinted, to seeing the mind as a genetic product with various innate capacities to respond and learn. The broadest and most fundamental view is that expressed by Plotkin (1994: ix), who states that, 'To know something is to incorporate the thing known into ourselves. Not literally, of course, but the knower is changed by knowledge, and that change represents, even if very indirectly, the thing known.' Plotkin goes on to develop the idea that in this sense knowledge is a fundamental aspect of adaptation, by which the living creature responds to the characteristics of the world it inhabits. He takes the view that 'We simply will not understand human rationality and intelligence, or human communication and culture, until we understand how these seemingly unnatural attributes are deeply rooted in human biology' (Plotkin, 1994: xiv).

This premise is pursued by considering how the neural circuits in the individual human brain produce selective adaptations in response to needs or problems, which parallels the selective adaptation of individuals in the species as a whole as proposed by Darwin. This is a foundation of what is now often referred to as neo-Darwinism or universal Darwinism, which is an extension of evolutionary theory. It is clear that these ideas will have to be considered at some point in relation to the neural computer, or the true thinking machine, which is postulated as the next great step in computer development. But for the present, this broad view of human knowledge is a sufficient starting point.

In the full sense, knowledge may involve knowing through the senses as well as knowing through the rational mind. Both are valid in ordinary human experience. A remembered fact or event may bring both elements into play. Even though most philosophers in the past distrusted information provided by the senses, there are many types of knowedge that are impressively effective, yet are obviously based on direct sensory experience. Distinguishing, recognizing and remembering human faces is a common example. Even after only a brief acquaintance, and sometimes after a long time interval, a brief glimpse may bring out the 'I know that face' response. Yet, as many people have pointed out, even though we 'know', we find it extremely difficult to describe human faces in an intelligible way, despite the fact that the records stored in the brain must be highly efficient.

The fact that a great deal of knowledge can be memorized, recalled, thought about and articulated, and that this is primarily the knowledge used in a great range of conscious and purposeful activity, should not be allowed to totally obscure the equally important fact that there are other ways of knowing which are functionally significant. In particular, visual perception is an important means of acquiring knowledge through the senses. Because both map making and map using depend on the visual system, both its internal organization and its relationship with intellectual and verbal responses are critically important in the making and the use of maps.

There is, however, a third aspect of knowing, which is often expressed as 'knowing how'. Yet it is quite different to what is normally meant by knowing. Human beings all acquire the ability to stand and walk, and in this sense know how to do these things. Although they are instinctive, they have to be learned, but they cannot be learned by formal instruction. The repeated attempts of the small child gradually succeed in establishing the required control. But no one can actually describe the actions that are necessary to stand or to walk. Significantly, it is these apparently basic actions that are most difficult for computers to deal with. Despite the presence of all sorts of 'artificial' creatures in science fiction films, those that walk are human beings in tin suits.

Skill

Apart from these instinctive behaviours, which are conditioned by the needs of the species as a whole, there are similar kinds of human performance which are deliberately learned. These are generally referred to as skills, although again the term is often misused. Many basic activities, such as writing, can be performed at a high rate and with little apparent effort, but only after a long period of practice. Significantly, they all depend to some degree on the senses, and therefore are a form of experiental learning. Almost any activity which involves the control of tools, especially the coordination of hand and eye, depends on skill. What is significant is that all such skills can only be acquired through active participation, frequent rehearsal and practice. No one can learn to ride a bicycle by reading about it, nor become a good tennis player by studying coaching manuals. Although they can be encouraged and assisted by organized practice, every person has to acquire skill individually. It is also the case that a person can know a great deal about something without necessarily having any skill, which is why artists are usually unimpressed by critics.

This is not to say that there is some sort of complete separation between skill and knowledge. Every skilful person also has knowledge about the medium, the activity and/or the tools involved. In fact, many skills are improved by consciously learning about factors which affect them. As skills are acquired individually, there are also considerable variations in performance by one individual, as well as differences between individuals. Therefore they overlap into talent, which is normally regarded as an individual capacity to acquire and perform some skill at a level beyond the ability of most other people. Again, this is a term with several shades of meaning. The *Shorter Oxford English Dictionary* definitions include 'A special natural ability or aptitude . . . a natural capacity for success in some department of mental or physical activity.' Some people seem to have a natural 'gift' (from the notion of talent as God-given) for playing a musical instrument, writing poetry, or even playing football. But whatever the propensity, it has to be developed by practice, like other skills. All the arts, whether creative or performing, have this requirement for skill, and in most cases such skills are only fully developed through a long period of constant attention and practice. These skills are not mindless routines – they are used to a given end, and in the case of the graphic arts, to accomplish what has been conceived in imagination.

In cartography, the obvious connection with skill is drawing, a point already commented upon. Does this mean that all other aspects of cartography involve only knowledge, especially intellectual abilities of the problem-solving kind? In many cases, something called experience is treated as important, but does this experience represent an accumulated knowledge of cartographic practices and methods, which could be learned from books, or does it involve another element which is much more closely associated with skill?

It has been explained that to make an adequate map the problem of adjusting level of information, scale and symbolic representation has to be solved, even though the solution is unlikely to achieve all the desirable objectives simultaneously. As soon as the outlined map content is displayed on the screen, even in the most rudimentary form, the experienced cartographer will respond visually to what is perceived. It is this visual reaction that may lead to the conscious mental response that 'something is wrong', enabling this to be treated as a problem to be solved. The non-cartographer, looking at the same image, is less likely to apprehend the cause of the problem. Solving the problem may involve the mental question, 'What will happen if I make the scale larger?' – a response which may be initially considered through visual imagination. Both these reactions make use of the cartographer's habit of visual attention, which has led to the accumulation of experience through practice. In many cases, the experienced cartographer will be able to identify a potential problem before any map construction takes place, by visually imagining the relative effects of the scale of the map and the proposed content and symbolization.

In these ways the cartographer obtains knowledge through the senses, but is also able to interconnect the thinking aspect of problem solving with images conceived through visual reaction or thoughts translated into visual images. It is not the case that cartographic problems of this sort are dealt with through some inexpressible and quite independent 'intuition', or some equally vague 'artistic ability'. Cartographers, like other people, do not analyse such procedures because it is unnecessary, nor do they comprehend what takes place in the brain as skilled performance. But it is this combination of skill and knowledge that is most difficult to describe, or to define by formal rules. It is the habit of visual attention, and the readiness to react to and to trust what is perceived visually, that characterizes both the skilled map maker and the skilled map user.

Even the briefest acquaintance with modern cartographic literature will show that it is precisely this kind of skill which is virtually ignored in many accounts of cartographic activity. This, no doubt, is largely because it does not fit tidily into any curriculum, and this in turn is influenced by the fact that visual learning is virtually ignored in educational systems that concentrate attention and effort almost entirely on words and numbers. True, there are graphs and diagrams and today sophisticated images, but they are always subordinate to language, and the concepts and processes that can be handled by language.

Artificial intelligence and cartographic communication

Given that much of the purpose of the earlier advocacy of using communication was as a model for understanding map making and map use, and that this model was ostensibly founded on the principle that the study of cartography must advance to the stage where it could be treated as a 'science', it was hardly surprising that the mechanistic model of communication and behaviourist theories of psychology were widely accepted. For some, therefore, the use of the digital computer to solve cartographic problems and replace the 'subjective' cartographer by a properly organized body of verified knowledge would appear to be a logical advance in this direction. True, map design is a problem, but if design is essentially problem solving, then machines that can 'think' as well as store and recall a vast amount of information should eventually be able to solve design problems as well.

Of course there was, and is, considerable opposition to the notion of an intelligent machine, for a variety of reasons. For some, it signals yet another step towards the domination of human society by the powerful, and the replacement of people by machines. For others, it undervalues or ignores the human requirement for aesthetic values and the fact that human thought and action is guided by intention, even if the intention is at a subconscious level. For some, at least, the gap between any machine and a human being appears to be so great that they see no real possibility of bridging it, a point forcefully expressed by Dreyfus and Dreyfus (1988: 39), who maintain that 'Intelligence has to be motivated by purposes in the organism and goals picked up by the organism from an ongoing culture.' More importantly, perhaps, those who have been engaged with artificial intelligence research for a long period can have considerable doubts as to whether it can be developed from its earlier promises. Waltz (1988: 192) makes the point that what is needed is an appropriate science of knowledge. He says, 'For some aspects of knowledge, any computational device will be on a strong footing compared with a person. . . . For such understanding to be deep, however, a system needs perceptual grounding and an understanding of the physical and social world.' He also observes that for the machine to become truly intelligent, and therefore capable of learning both directly and indirectly, it will have to be provided with adequate sensory systems, enabling it to learn from experience.

To some degree these cautionary responses are a consequence of an early concentration within artificial intelligence research on physical symbol systems and heuristic search theory, which, Waltz (1988) maintains, were consequent on digital computers and the psychological theories of behaviour current at the time.

Boden (1987) makes a sustained and powerful argument for a quite different point of view. She maintains that it is in the process of trying to make machines that have some of the intellectual capacities of human beings that it becomes even more necessary to learn more about what constitutes human intelligence in the widest sense. She says (in the preface to the second edition), 'Contrary to what most people assume, this field of research has a potential for counteracting the dehumanizing influence of natural science, for suggesting solutions to many traditional problems in the philosophy of mind, and for illuminating the hidden complexities of human thinking and personal psychology.' In this sense, the desire to program machines to be intelligent means that the sheer complexity of the human mind has to be accepted, because any success, however limited, can only be achieved by gradually improving and extending our knowledge of how the mind works in its fullest sense. Despite the limitations of the present systems, therefore, they should be regarded as attempts to improve our understanding of human intelligence, and this in turn can be enhanced by attempting to model it artificially.

Cartographic generalization

In the earlier discussion of generalization, the emphasis was placed on its effect on the map from the user's point of view. Before going on to consider how generalization can be dealt with in a digital mapping system, it seems to be desirable to describe generalization as traditionally practised by cartographers, for it is this collection of operations which the digital system needs to replace.

Generalization in itself is not a cartographic objective. A cartographer does not set out to 'do' generalization. The first stage in producing a map that can be derived, wholly or in part, from existing material is to decide the scale and nature of the desired map. The compilation of the map, assuming that it follows the usual practice of working from more detailed sources, will involve generalization. In many cases, especially at very small scales, the first problem will be to find suitable source material at an appropriate scale. For small-scale maps of large areas, this is likely to require the use of more than one map, and the sources are unlikely to be homogeneous in terms of scale, coordinate systems, projection and content classification. Where a small-scale map is derived from a single map source, and the map to be produced can be based directly on it, the generalization task is much easier. Significantly, all the experimental work carried out so far falls into the latter category.

The difference in scale between source map or maps and derived map is significant. Few cartographers would attempt to carry out a generalization where the scale difference is more than 3 : 1. Using analogue methods, the normal practice is either to photographically reduce the source map to the desired scale, and then produce the generalized compilation on this basis, or to produce the initial generalization at the scale of the source map and then reduce this. In this case, it is axiomatic that the line symbol specifications of the intended map are enlarged to the scale of the source map, so that the resultant generalization will be appropriate to the final scale. With a greater scale difference, the common practice is to reduce the source map to an intermediate scale, and enlarge the symbol specifications to match this.

Differences in scale introduce more problems than just numerical ratios. With topographic map series, for example, there are significant differences in classification and symbolization between large-scale plans and large-scale topographic map series. Large-scale plans of urban areas, at scales such as 1 : 1000, are not treated as a source for large-scale topographic series covering a whole country. In many countries, the largest scale at which the whole country is covered, and therefore the scale that provides consistent material for deriving smaller-scale maps, is at 1 : 20 000 or 1 : 25 000. Although in Britain the largest scale of complete coverage is the 1 : 10 000, in fact this is reduced directly to the 1 : 25 000 scale. (It can be argued, of course, that in this particular case, the 1 : 10 000 series is really an enlarged 1 : 25 000.)

Although further series can be produced by generalizing the preceding scale, problems arise if there is a considerable gap between one scale and another. In Britain, for example, where there has been no tradition of a 1 : 100 000 scale, the difference between 1 : 50 000 and 1 : 250 000 is considerable. A linear scale ratio of 5 : 1 is, of course, an areal scale ratio of 25 : 1.

The other factor, and one which tends to pass unnoticed, is that what is regarded as a suitable level of generalization for a map at a given scale is not a constant. For example, a comparison of small-scale maps produced in the former Soviet Union, and its neighbouring territories, makes clear that these are much less generalized – that is, they retain far more detail – than corresponding maps produced in the 'West'. In both cases, what is regarded as appropriate is a function of familiarity. The reason, presumably, is that in the former Soviet territories, only small-scale maps were publicly available. Users could not refer to larger-scale sources.

Generalization as a cartographic task

Once the problems of source material and scale of generalization have been sorted out, the task of compiling the new map or map sheets can begin. The central issue is to describe what the cartographer actually does. Most of the accounts which appear to be used in digital research tend to concentrate on generalization methods, or operations, but in reality these play a subordinate role. What the cartographer is trying to do is to provide a legible, coherent version appropriate to the smaller scale. This is affected by two sets of factors (see Keates, 1989: 38): quantitative factors, i.e. the scale difference between source map and derived map, and the relative sizes of the symbols on the two maps; and qualitative factors, i.e. the need to retain the characteristics of the topographic features and judge their relative importance. The scale difference between the two maps, and the space taken by symbols at the reduced scale, are quantifiable differences which are visually obvious. For example, the gauge of a standard contour line on the proposed map can be enlarged by the linear scale difference, and the width of this line at the source scale can be drawn. But considering the characteristics of the topography and judging the relative importance of individual features in the same class, or the relative importance of different classes, is a matter of interpretation and judgement. Most of this is done directly by constantly examining the detailed source map, not only in terms of individual features, but of the topography of larger areas. This is not a straightforward task, and in many cases could be improved by finding and using additional, non-map information, to help in making decisions.

Working with these objectives in mind, the cartographer constructs the new version at the reduced scale. The procedures involve eliminating whole classes, selective omission of some features, simplification of lines and outlines, and frequently combination, exaggeration and displacement. For a topographic map, these operations are always carried out in a given order, because of the interrelationships between physical and cultural features. The order is coastlines and shorelines, drainage systems, contours, settlements and communications (which have to be treated together), boundaries, vegetation and land use. As cultural features occupy specific positions in relation to the physical land surface, their generalization has to follow any changes made to the physical features. If a road follows the bank of a river it must retain this location in the generalized version. If a house is located at the side of a road, it must stay in this position if the road is represented by a relatively larger symbol at the smaller scale.

These are the problems that the cartographer has in mind when carrying out the generalization: her attention is focused on objectives, not procedures. In most cases, operations such as simplification, exaggeration and displacement are carried out simultaneously. In dealing with a built-up area, for example, outlines are simplified, small enclosed spaces are omitted, major street lines displaced, and groups of adjacent buildings are combined, and all this happens as a single operation. The idea that generalization is based on 'knowing' about individual processes is quite false. (If this seems to be strange assertion, I can point out that as a young cartographic editor, I compiled many small-scale maps before anyone told me that what I was doing involved something called generalization.)

The point to be emphasized is that the practice of cartographic generalization is essentially a visual skill. Constantly examining the detail of the source map, moving from peripheral vision to central vision and back again, understanding what the symbols represent as topographic features rather than treating them as 'lines' or 'points': these are the typical processes. In analogue production, the act of actually having to draw the generalized version also reinforces an understanding of characteristic forms, even though it exists at a subconscious level. Cartographers become familiar with different coastal forms, the smooth curves of sandy beaches compared with the sharp irregularities of headlands and estuaries, as hand and eye work together. It requires the same understanding of the connections between the immediate detail and the imagined whole which is characteristic of the graphic arts, like painting. Whilst looking at the painting as a whole requires a broad view, the placing of each individual brush stroke requires concentration on detail. The eye and the imagination constantly move from one to the other.

Now it can be argued that the simplification of each individual line will be more objective and more consistent if processed by the computer than if left to a cartographer. But precise simplification is not the total objective. What matters is a correct and intelligible representation of the relief forms and topographic characteristics of the area at a smaller scale. It is in such matters that cartography lies closer to art than science: *'L'exactitude n'est pas la vérité'* (attributed to Matisse). Even apparently routine tasks in cartography can only be carried out if the cartographer understands the function of a particular element in the map as a whole, and is aware that the relationship between what the map states and the external world is contrived, through methods of representation and symbol design, to a given end. Such understanding requires intelligence, but it is an intelligence which often keeps 'rules' at a subconscious level.

Research in generalization

Attempts to carry out at least some types of generalization by computer processing have been made over a long period. For example, Pannekoek (1962) experimented with simplification of coastlines and contours. With the development not only of much more sophisticated cartographic databases, but also geographical information systems, the generalization problem has been attacked with renewed vigour in the last decade, and an impressive amount of research has examined generalization theory and practice.

Broadly speaking, this research falls into two main groups: research aimed at developing or improving a specific generalization process, such as line simplification or selective omission; and research aimed at developing a complete generalization system capable of performing all the tasks involved in producing a derived map. The first can be exemplified by Buttenfield (1991) and Visvalingam and Whyatt (1993), and the latter by Brassel and Weibel (1988). Whereas the former type of research seeks to obtain an equivalent result to manual generalization, but often by quite different means, the latter is closer to the expert system approach as it depends fundamentally on using the 'knowledge' of the cartographer as a basis for the system.

Whereas the algorithms used for independent line simplification will give a predictable result under specific conditions, and can be modified to provide different degrees of line 'smoothing', attempts to solve more complex generalization tasks have all relied on finding out what 'rules' cartographers use. The assumption is frequently made that generalization depends on knowledge which in turn must be used to devise a set of rules. McMaster (1991: 38) states this plainly: 'It is clear that for the successful implementation of rule bases, and ultimately for fully operational expert systems, appropriate models of generalization must be designed, a comprehensive set of generalization operators must be established, and, perhaps most importantly, cartographers must identify and logically encode "rules" for the process.'

Extracting the rules from the expert is of course the common basis of all expert systems, and a common cause of complaint by knowledge engineers. The source of expertise can be either what has been published about generalization in books and journals, or what can be obtained by consulting cartographers. In an interesting attempt to analyse this cartographic expertise, Rieger and Coulson (1993: 71) asked 23 cartographers to 'consider the following domain: any and all generalization procedures (tasks, steps) which they could use to alter map data,' and went on to state that 'Declarative, factual knowledge about procedures is being sought, not procedural knowledge, i.e. how the generalization task is carried out.'

Given that the answers were consequent on the nature of the question, what they discovered was that the cartographers did not describe their 'knowledge' very well, did not use terms the same way, did not agree with each other, and that there did not seem to be any common understanding of most generalization processes. Similar problems are reported when published descriptions of generalization are compared, in so far as they deal with the so-called generalization processes. The obvious conclusion to be reached from this and other evidence is that perhaps the rules are not of major significance to a cartographer, and any knowledge the cartographer may have is held at a subconscious level. It is well expressed by Mackaness (1991: 217), in saying that 'The classification of generalization techniques is an academic process and the cartographer is not consciously aware of such distinctions when designing a map.'

In an interesting comparison of manual generalization and digital processes, Nickerson (1991) uses the production of Canadian 1 : 250 000 scale topographic map sheets from 1 : 50 000 scale source maps. Some of the sheets concerned included complex areas of terrain dominated by small bodies of water and watercourses typical of the Canadian Shield. The manual generalization was carried out at an intermediate scale of 1 : 125 000, the symbol dimensions being doubled to allow for final reduction. The 'rules' which seemed to preoccupy the cartographers, according to Nickerson, were concentrated on keeping character both for individual features and areas, deciding priorities in displacement, and maintaining requirements for graphic legibility such as minimum line separation. These are of course concerned with objectives, not simply the application of processes. The main conclusion is that 'successful cartographic generalization relies on visualizing all of the data in an area so that generalization decisions can be made' (Nickerson, 1991: 53). The central problem is of course that it is impossible for the digital database to visualize all of the data in an area simultaneously.

It is clear that many researchers have realized that any rule-based approach is unlikely to provide a complete answer, and have appreciated that generalization is not simply a technique but a visual skill. Beard (1991) refers to 'individual artistic skill'; Mackaness (1991: 218) concludes that 'the human eye coupled with the artistic cartographic hand provides an immediate solution.' The alternative possibilities seem to be either to adopt an entirely different approach to computing, or to attempt a combination of digital processing and human control. The former is advocated, for example, by Langran (1991: 207), who proposes that 'parallel rather than sequential models of information processing seem more likely to mimic the human decision process,' but of course this is, at present, only at a theoretical stage. The latter view is most clearly expressed by Weibel (1991) in an approach described as 'amplified intelligence', and is based on earlier work by Brassel (1985) and Brassel and Weibel (1988). Weibel (1991: 186) concludes that 'The application of expert systems strategies to map generalization has not yet been very successful. The lack of performance is mostly related to limited effectiveness of knowledge engineering.' He goes on, 'Amplified intelligence systems take an intermediate position between algorithmic and pure expert system strategies. They can initially be based on existing techniques (algorithmic or knowledge-based methods) that solve individual parts of generalization.

The guiding knowledge ensuring proper application of those procedures is contributed by a human operator (expert).'

This proposal, which is explained in detail by Weibel (1991), is even more interesting because it suggests that instead of regarding 'manual' and 'automatic' as fundamentally different, and in some senses opposed, the combination of both approaches would be valuable.

In adopting this idea it is also possible to recognize a fact which itself is rarely mentioned: that generalisation by cartographers is far from perfect. In map compilation, generalization requires a high level of concentration. Like many aspects of map production, it is frequently done under pressure, as the time allocated for any part of the production process is limited. Minor errors are difficult for the cartographer to detect, because of course the generalization is an interpretation, and there is no other 'model' with which it can be compared. Line simplification algorithms could not only be implemented to give a consistent result, but variations in the degree of smoothing could be tried and compared. Procedures for selective omission could be used for the rapid production of a given element, and then checked by the cartographer to see if they have given an appropriate result. If much of the drudgery could be removed (and cartographic generalization can be a tedious task) then cartographic skill could be concentrated on solving the more complex problems. Regarding machine and human operator as allies, rather than rivals, is a far more promising approach.

Weibel (1991: 186) also suggests that by being more modest in the original approach, 'It is hoped that by analysing how cartographers use the system's function, expert knowledge can be formalized in terms of automated methods and built into the system.' In this sense, the interaction of the cartographer and the system, not simply by theoretical studies, but through the actual execution of generalization for specific maps, is more likely to lead to the eventual construction of a fully automatic system. Significantly, the approach described by Weibel has been adopted by Intergraph as a basis for their Map Generalizer system (Lee, 1992). It seems that in the system, 'Dynamic display (per feature(s)) of the generalized result is available. This helps users to visually compare the results and optimize the parameters' (Lee, 1992: 8). If this generalization can be viewed both at the source scale (with enlarged symbols for the generalization) and at the reduced scale for the final version, then it will come very close to imitating the common procedure of cartographic practice.

The search for a process of cartographic generalization is in its own way an interesting example of what Boden (1987) was referring to when describing the prospective advantages of trying to model human intelligence. Sustained research leads not only to a better understanding of what computer processing can do, but also to a recognition of the remarkable capacity of the human brain to selectively summon what is relevant and consider several factors simultaneously. In turn, it should help to remove the notion, unfortunately still prevalent, that the 'art' in cartography is some sort of decorative addition rather than a fundamental characteristic of cartographic work.

Name arrangement

Much of the research published on name arrangement suffers from the same weaknesses as that on generalization. The early attempts concentrated on the simplest aspect – the positioning of names related to points or small areas treated as points – at relatively large scales, which rather minimizes the problem. On medium- and large-scale maps, that is 1 : 50 000 and larger, names occupy a smaller proportion of the total map space than they do on small scales. Therefore they are less likely to intrude into major map detail, and the space available for alternative positioning is greater.

Once again, there is a preoccupation with rules and variations on rules which is quite irrelevant in practice. The objective of a proper name arrangement is to aid legibility and ensure correct identification of the named place or feature, and to do so causing minimal interference with other map information. The so-called 'rules' are a guide to this, but in fact they are quite simple (see Keates, 1989: 116–119). The major problem with name arrangement is not the 'ideal' placing of the names themselves but the geographic circumstances and characteristics represented by the other map information. There are no absolute rules because the desirable objectives are often incompatible. Legibility and ease of reading suggest that names should be placed at or near the 'horizontal' or normal position of printed text. But correct identification of a mountain range which runs more or less north–south means that the name should be aligned along the major axis. In practice, correct identification takes priority over ease of reading, but technically it is far easier to place names 'horizontally' than to arrange them on an angled curve.

Every part of the Earth's surface is unique in the sense of its particular topographic detail. The alignment of the coastline, the presence or absence of rivers, the concentration of cities in small regions, and the length of the names are all factors which vary from place to place. Name arrangement is obviously most difficult at small scales, where names occupy a large amount of space relative to other features, and can obliterate or interrupt other map information.

In traditional (analogue) cartographic practice, the name arrangement is normally the last task in compiling a new map, because the placing of the names is conditional on the rest of the map information. It is often most critical where a group of towns and cities lie close together, and where there are other constraints such as coastlines, offshore islands and international boundaries. On very small-scale maps, the problem is so acute that the decision to include a town or city will be directly influenced by the possibility of including the name. Just as in generalization, it is the conditions that exist in a given area, not simply the connection between a name and a feature, which are of paramount importance, and which the cartographer tries to solve by considering the possible positions of all the names in the area. Once again, it is essentially a visual activity, in which the eyes move from the larger area to the detail of the exact positioning of a name, or even individual letters of the name. This does not require factual knowledge. The factors that influence the placing of the names are evident by looking at the map compilation and being able to understand their implications for the particular case. It is necessary for the cartographer to know the typographic style and dimensions of each name, and therefore the space it will occupy when placed on the map. This also requires visual imagination, and like generalization, it is easier to carry out satisfactorily if the cartographer has actually produced name arrangements manually, where the imitation of type size is necessary. What is important in this is not drawing skill, but visual attention to detail and visual imagination.

The problems and preferred practices of name arrangement have been explained and demonstrated in many texts (see Cuenin, 1972: vol. 1, 241–245). In dealing with names attached to point features, the requirements of the names of areas, whether physical or administrative, have to be considered, and often it is these names, with their larger spaced lettering, which are placed first. Similarly, if there are few good possibilities for the placing of a river name, the available space will be kept clear of other names which might have occupied it. Because the cartographer can see the whole map compilation, the possible or probable placing of adjacent point names is taken into account. For example, in adding the name of a mountain range, the space required for named peaks and heights is taken into consideration.

In congested areas, the best solution is not always found at the first attempt, and name arrangement must be one of the few cartographic processes that can at times involve an iterative procedure. Few cartographers who have done a great deal of name arrangement can claim that they always got it right first time! But the ability to realize where the possible spaces lie, and to translate this into an arrangement in accordance with the other map detail, is essentially dependent on the use of both central and peripheral vision, moving from the general to the particular as required. It is this combination of visual perception and an understanding of the objectives, as well as the 'knowledge' of rules, which sustains cartographic skill. As researchers constantly rediscover, cartographic practice is often intelligent activity of a high order, not just a collection of 'techniques'.

CHAPTER 20

Expert systems and map production

Although artificial intelligence has led to the use of computers to imitate human thought and actions in many fields, such as robotics, the developments currently receiving a good deal of attention are generally known as expert systems, or more correctly, knowledge-based expert systems. These lie in a sort of half-way house, for the underlying premise is that the system will make use of the knowledge of a human 'expert' in order to arrive at a set of rules for dealing with a particular problem area (domain), rather than a truly intelligent machine system that can identify the problem, learn whatever it needs to know, and provide the answer or product. On the other hand, an expert system is expected to do more than, for example, computer-aided design, in which the computer merely aids the operator by rapidly producing and changing images according to the operator's instructions. The aim of the expert system is to make possible the production of a finished map from a given database, which in a sense is more ambitious that the 'intelligent' programs for generalization processes or name arrangement, which are limited to specific tasks within map production as a whole. Given that all the cartographic operations needed to make a map must be included, it must be more complex than dealing with individual processes. On the other hand, it is quite clear that at least to begin with, any expert system can only be expected to deal with a very limited range of maps at a very simple level.

Before considering the ways in which this basic idea is being applied in cartography and GIS, it would seem appropriate to enquire into the reasons for this interest. In some ways they may seem to be obvious, but apart from the usual ostensive reasons, there are a number of assumptions which are rarely stated. Without assigning any significance to the order, they can be listed as follows:

1. because mapping capabilities are available to users with little or no knowledge of cartography, who therefore need guidance, especially in design (see Green, 1993);
2. because cartographers (using the term in the broadest possible

way) are too expensive for 'the market', and must be replaced as far as possible by more efficient machines (see Orr, 1991);

3. because of a general dissatisfaction with cartography and cartographers, as cartographic practice seems to be 'unscientific';
4. because cartographic theory and practice can be improved by trying to develop computer systems based on proper research into cartographic tasks; and
5. because research of this type seems appropriate in an academic environment.

The last three overlap, but this is not meant to imply that much of such research is not the result of a genuine interest in the problems concerned. It is simply that the circumstances which cause people to act in a certain way are affected by wider and usually unspoken assumptions current in a given society at a particular time, and the general drive to be keeping up with the progress of technology is one of these.

Artificial intelligence and expert systems

Reeke and Edelman (1988: 149) state that expert systems 'combine rule-based inference and natural-language interface techniques with domain-specific databases.' They point out that such programs, however efficient they are in dealing with logical problem solving, are 'insensitive to context and are likely to give quite incorrect responses to queries that are slightly outside the domains for which they were programmed.' Waltz (1988) draws attention to the dependence on physical symbol systems and heuristic search as basic elements which reflect the characteristics of serial digital computers. He makes the point that if a system is to be able to learn, then a central question is how knowledge should be represented, which in turn implies a far greater understanding than we presently have of how knowledge is structured in the brain. As Katz (1994) points out, 'The great paradox of the emerging field of Artificial Intelligence is that computer scientists found that the easiest aspects of human behaviour to get computers to emulate were the ones we often considered most difficult while it has been the everyday intellectual skills we take for granted – what you might call basic common sense – which have eluded them.'

The idea that a system should be able to learn is frequently suggested as a desirable objective – indeed *the* objective – even though the building of the system has to start by applying knowledge obtained from a human expert. Theoretically, at least, the operation of the system, dealing with a variety of tasks, would, on the one hand, enable its designer to improve its performance, but on the other hand, the system itself should gradually be able to build in the capacity to solve more problems without human intervention. This, eventually, would make it truly 'intelligent'.

Expert systems in cartography

The basic components of an expert system as currently considered to be applicable in cartography are described by Forrest (1993). They contain a database, a knowledge base, an inference engine and a user interface. For operations to be carried out by processing the data, the system requires structured knowledge and also a means of solving problems. This knowledge base contains facts about the data and rules for operating on it, supported by procedures for inference and problem solving. According to Ranzinger (1985: 183), 'In expert systems this knowledge is usually handled explicitly as a separate entity named the knowledge base and not contained implicitly as part of the program code.'

The nature of the knowledge base raises several questions in itself. Is knowledge in an expert system simply a set of definitions or descriptions, so that the system 'knows' what a contour or a point symbol is? Or does it need to encompass a much broader range of knowledge of the external world? For example, the cartographer understands the relationship between the shapes shown by contours and what they represent in terms of hills and valleys. This in turn is related to any drainage system, so that the quite separate sets of symbolized information have to be compatible. So is cartographic knowledge or expertise simply a knowledge of representational methods and types of symbol, or does it need the ability to relate these to the characteristics of the phenomena being represented?

Theoretically, the system should be able to understand a request for the production of a certain type of map (within the limits of its database) and then itself take the necessary steps to produce it, by guiding the system user through the stages of production. Therefore it must be able to interact with the user, because in many cases several different possibilities will exist and different users will have different requirements. At the same time, it should have sufficient knowledge of the domain and possible outputs to prevent the user from making wrong, or at least catastrophic, decisions.

Typical of inference operations are the processes known as forward and backward chaining. In the former, rules are applied to define the problem as determined by the user, and the system searches for a solution to the problem. In backward chaining, the ultimate requirement (in this case a particular map) as proposed by the user would be checked against the facts to see if it is possible to produce an appropriate solution that would be in agreement with accepted cartographic practice. If this did not arise initially, then another possible solution would be selected and checked.

Although such methods can operate on factual analysis, where there is a specific answer, there are many aspects of design that are normally considered only to the point where an adequate solution occurs, certainly not the only possible one, and not necessarily the best one. This is commonly referred to as the area of heuristics, which in this sense refers to the fact that the human being (especially the expert) seems to choose an initial path to explore without consciously searching through rules at all. In human terms the method is by no means perfect, but most of the time it is at least adequate, and it avoids storing and processing the large amounts of data and sets of rules that are needed for serial search.

The fact that the 'expert' does not necessarily arrive at a solution, or even a possible solution, by any apparently logical inspection of rules, but simply seems to 'know', applies to many types of expert system, but in particular where the human expertise involves visual perception. As Schwartz (1988: 124) observes, 'Since extensive serial symbolic search operations of this type do not seem to characterize the functioning of the senses, the assumption (typical of much of the AI-inspired cognitive science speculation of the 1960–80 period) that serial search underlies various higher cognitive functions becomes suspect.' It was in this sense that Michie (1989) complained that it was skilled behaviour that computers needed, not simply more knowledge.

These general observations are important in an examination of actual or possible cartographic developments. If, as has been proposed here, an important part of cartographic practice involves skilled behaviour, and in particular the interlinking of visual skill and associative thinking, then enabling a computer to deal directly with map design is likely to be a complicated task. As Petchenik (1977: 127) pointed out, 'If...we know a great deal more than we can tell, then there may be absolute limits on what we can know in a form that can be conveyed to the computer.' For those tasks where the eventual output is a list or table, then only the intellectual meaning of the solution is important. But for maps, the final output has a visual component which has to be related to representational meaning. From this it should be clear that initial attempts to produce cartographic expert systems would need to deal with very simple map types to begin with. Even so, it is at least desirable to consider what may be meant by 'design' and 'expert' in this context.

Design

It has been pointed out already that the term 'design' has two principal meanings. It can be used to refer very broadly to the general plan or scheme for a work, as in designing a new layout for a garden; or in the artistic sense, which refers to the giving of visible form. Whereas the overall plan for a map, embracing its purpose, geographical area, subject matter, format, scale, and representational methods, is essentially composition, symbol design more closely approaches this concept of the giving of visible form by specifying symbols, involving the graphic elements (point, line and area) and the graphic variables (form, dimension and colour).

The idea of planning the whole work, which is of course necessary, raises some interesting questions. If the final appearance of the map is conceived by the cartographer's imagination, then the preceding stages, within limits, could be seen as a series of connected operations devised to achieve this goal. Such a certainty of prediction would seem to be unlikely if the subject matter of the map is complex. Although a general concept of the map may be imagined and anticipated by the cartographer, making use of experience, it is at least more probable that this initial concept will need refining and modifying as the preparation of the work progresses. For the inexpert, such anticipation is highly unlikely, and indeed it is clearly one of the functions of the expert system to anticipate problems that are not realized by the system user.

Types of design

In practice it is possible to distinguish three major types of cartographic design: accidental design, design by imitation, and creative design. Many maps have been, and are, produced by people who are ignorant of, or simply indifferent to, cartographic principles of any sort. There are many maps, in books and the media in general, which show that the idea that a map should be treated as a complete composition has not even arisen. Certain lines and outlines are traced or copied, a few names added, and that is the map. Such apparitions can even be made more sophisticated by adding splashes of colour or elaborate typefaces, especially if the map is being made in an advertising studio along with other graphics. There is no control over cartographic incompetence, but one could wonder at what the reaction would be if newspapers and journals were written by illiterate journalists.

Design by imitation can operate on different levels. There are many types of map – and indeed some topographic map series demonstrate this clearly – where well-founded practices have developed into conventions. They exist because they work. Such conventions normally allow a good deal of freedom in individual expression, so that for example although generally water is symbolized by being in blue, the detailed choice of hue and lightness value can be adjusted to the conditions of the particular map. For nautical charts it is the users who insist on the maintenance of conventions as a basis for speed and confidence in interpretation and use. As has been pointed out previously, for many commercial companies, once a given style has been established, it is deliberately maintained, partly because of the cost of altering the visual appearance of a large body of material, and partly because it helps to identify the company's product in comparison with others. Therefore, there are many factors in preferring designs which maintain styles or conventions.

Despite the restrictions of imitation, the concept of design is also associated with the idea of making something that is new, which suggests a quality of creativity or originality. Such newness – which is normally justified as appealing to the consumer – may be quite superficial. The advent of 'designer' fashion is intended to suggest something original (and therefore necessarily expensive). Indeed the endless variations of fashion have now reached the point where the most important thing is not the value of the product itself but simply the name of the designer. But the illusion that such people are trying to sell is that original creative design exists, that it is the product of exceptional people, that it is intrinsically valuable, and that the outcome of such design is something unique. This happens because design is founded in art, and all art is devoted to the creation of original and unique things.

It is not easy to find convincing examples of truly original design in cartography. There are many reasons for this. An obvious one is that although the subject matter of maps may be as wide as the whole field of enquiry into the external world, methods of cartographic representation are limited. The same basic device is employed for many different purposes. Originality in design is more commonly perceived as an interesting or unusual combination of methods of representation and symbolic form, rather than a radical new departure. To take an example, there are numerous road maps and atlases of Europe and European regions and countries, some produced by national agencies, some by cartographic publishers, and yet others by tourist organizations, motorists' associations and so on. There are usually many alternatives for any region or country. Comparison of map sheets or atlases which treat the same region at the same or similar scales reveals endless variations on what is essentially common subject matter. They are all unique in the sense that they are all different, but they often differ more in details of treatment than general appearance. It would be difficult to maintain that the design of any one of them had taken place without any knowledge or cognisance of others. But it is also true that the cartographers who produce such maps and atlases are using their visual imagination and creative ability, and are not just following established practices.

So far as expert systems are concerned, the objective in assisting a non-cartographer to produce a map is to achieve minimum competence and legibility, not anything as intangible as creative design. Hence 'established cartographic practices' are frequently referred to, although it is often difficult to decide whether these are methods of representation or the giving of form. It may be that at some future time sophisticated expert systems may be useful in digital cartography by suggesting unconsidered possibilities, but this would still require a creative cartographer to make use of them.

Cartographic expertise

According to the *Shorter Oxford English Dictionary*, 'expert' is defined as 'experienced in . . . trained by practice, skilled . . . . One whose special knowledge or skill causes him to be an authority.' The root of the word is of course the same as that of 'experience'. Both knowledge and skill are included, but the notion of authority is central to the definition.

It is clear that anyone with specialized knowledge or ability is an 'expert' compared with others, but this is not the principal usage. An expert is normally regarded as one who exceeds the level of performance of other practitioners in the same field, this exceptional ability demonstrated by being able to deal with more profound or difficult problems. Although, like status, it is a condition that lies in the minds of other people, rather than a self-evaluation, its foundation is always dependent on a long period of direct involvement from which the experience is derived. It is not just experience but the ablity to learn from that experience, raising performance to a higher level, that is critical. Like skill, it is not a faculty that can be acquired second-hand, because it is the product of active participation.

In the cartographic literature on expert systems in general, there is surprisingly little examination of human experts. Given that such expertise is demonstrated by maps, rather than publications about cartography, seeking it out raises difficulties for people operating mainly within academic or research fields. One intriguing consequence is that the 'expert' tends to be regarded as whoever published the first account of a practice. For example, in dealing with name arrangement, many people refer to Imhof (1962) as an expert source. Yet the principles so described were taught to me in 1950 by a former surveyor general of India, who explained that they were common knowledge in the Survey of India cartographic drawing offices in the 1920s.

Contrary to what Müller and Wang (1990: 25) suggest, it is not a 'substantial body of research which has dealt with the principles of cartographic symbology and design' but the accumulated knowledge and understanding developed by practising cartographers over a long period. All the 'techniques' involved in specialized maps, including those based on statistical data, such as dot distributions, proportional symbols, graded series in colour or monochrome, quantities applied to unit squares, flow lines and so on, were well-established practices before any of the modern 'research' into cartography came into existence. Significantly, these practices, unlike a good deal of theoretical research, always recognized the limitations of visual perception. They are well exemplified in the European national atlases, produced by analogue methods.

It so happens that this evidence for cartographic expertise comes from maps, which in terms of cartographic design can rarely be linked to individuals. It is also the case that this experience is unevenly distributed within the 'cartographic' community. Although references to 'cartographers' and 'cartographic experts' in the expert system literature are generally vague, in Europe at least there are significant differences in the range of skill and knowledge acquired by individuals within map production as a whole. In 'traditional' cartographic production, they are most developed by people who are not called cartographers at all, but map editors or cartographic editors. In many European languages, the connections between map, map making, map drawing and map editing are all obvious because they all use the same root, *kart*. For most of the specialized cartographic companies – which normally do cartographic work for a range of clients, as well as their own publications – it is the cartographic editors who deal with clients or proposals for new publications, design the prototypes, plan the work, compile and design the maps, and supply the necessary information for the production schedule and cost estimate. The chief 'cartographer' is normally responsible for maintaining graphic standards, organizing the production staff, and especially in costing and estimating the manual drawing processes.

In many cases, where the organization deals with a great range of scales and map types, the chief problems lie not in methods of cartographic representation themselves, but in creating useful products from a variety of sources within limits of time and cost. Alternative methods of representation are discussed, but always in terms of the map as a whole, because given an outline of the proposed specification, the cartographic editors are conscious of what they would look like in the particular map. In such circles, the 'expert' is not just someone who can deal with such tasks as generalization (which is taken for granted), but one who can devise a visually satisfying solution within all the constraints externally imposed by the publication or the client. The constant need to generate new designs for different types of map is the testing ground for the development of 'expertise'. What this produces is not a conscious knowledge of 'rules', but the ability to apply embedded experience and skill to new tasks.

In other types of map-making organization, especially those dealing with large-scale plans and topographic maps, new design or content specifications are rare, and consequently cartographers are essentially those who deal with production. Most of the staff at management level are engineers (surveying engineers, topographic engineers, cartographic engineers or even geographical engineers) and they are well aware of the connections between content and representation. Designs for new maps or new series are examined and tested at length. Because so many of them are also experienced map users (especially the military engineers), they bring a critical eye to cartographic design.

Given an understanding of the new technology, experienced cartographic editors can employ their expertise in the new technical environment, and at least provisionally can continue to deal with tasks such as generalization and name arrangement on this basis. To what extent such relatively slow output, normally interactive, will be replaced by 'intelligent' programs of the type referred to previously, remains to be seen.

Knowledge, rules and procedures

In developing an expert system, therefore, there are two related sets of problems. The first is to transform existing cartographic practice into rule-based knowledge, and the second is to guide a non-expert through the map-making task. Provided that the former is sufficiently limited, the latter may become possible, in that the system can provide a series of menus which ask the critical questions, and then suggest or demand specific answers or solutions. Müller and Wang (1990: 25) take this to its logical conclusion by omitting the term 'expert' altogether, on the basis that 'At the present time, most expert systems should be more accurately called knowledge systems because they can not perform at the level of a human expert.' Whether this accepts that there is a difference between 'knowledge' and 'expertise' is not made clear.

In this area there are obvious similarities between expert systems and the type of artificial intelligence applied to map generalization. Cartographic knowledge is realized through specific applications. When asked about making a particular map, the cartographer consciously goes through a set of questions in order to determine the characteristics of the proposed map, and to resolve any conflicts between requirements. The problem of leading the inexpert user through the processes needed to produce a satisfactory map seems to be treated as though it is an entirely new departure in cartography, but in fact it is something that generations of cartographic editors have had to deal with. It can be illustrated by a simple, if imaginary, example.

Suppose that a prospective client asked a cartographic company about the possibility of producing a map. The client is the sales manager of a large company who has been asked to write an article on new developments for the company's house magazine. He regards the opening of new retail outlets as a major theme, and thinks that a map showing these might be helpful to his article. Having established this, the cartographic 'manager' would then have to elicit what exactly was wanted in cartographic terms.

A series of questions would then reveal that most of the company's outlets were in the south of England; that there were about 200 of them, and that in the last two years 15 new ones had been opened, a point which the author naturally wished to stress. Further questioning would establish that these outlets were in fact graded: a few large ones in major cities, medium-size ones in larger towns, and a majority of small ones in small towns and villages. The client's house magazine was briefly examined: standard A4, black with an additional orange for 'design' purposes.

By this time the geographical area and the format are known. The client does not think a two-page spread should be necessary, so the overall size of the map is fixed. What else might be wanted? 'Do you want the names of the places where the retail outlets are located?' After some discussion it is agreed that the names of major city outlets and large town centres should be shown, and the client is brought to understand that naming all 200 locations would be impracticable. Despite being asked about county boundaries, relief, communications, etc., he is quite clear that he does not want anything else on the map.

Two other points are immediately clear to the cartographic manager. Locating small towns will probably mean going to 1 : 250 000 scale sheets for identification. Emphasizing the new outlets can be done quite easily by reinforcing whatever symbol is used, or even using a different symbol, such as a star. If the locational symbols, suitably graded in size, are shown in orange, the names can be in black and the land area brought out by showing the sea in a weak tint of the black. That leaves the title of the map and the company's logo. In practice there will probably be more discussion (and changing of mind) about this than about the map itself.

In this case, the cartographer knows that if every location on the map were to be named, many centres would be obliterated by names, and the names would attract more visual attention than the location symbols. But if the names were shown in a colour light enough not to interfere with the symbols, the primary information, then they would not be legible.

Now of course this is not an exact parallel with the problem of guiding an inexpert user to make the map without the help of cartographers, and obviously the client in this case does not have to make any cartographic decision independently. Nevertheless, it demonstrates that putting the map together starts by determining the overall composition, and that questions of symbol design arise from this. Far from being the major difficulty, once the other questions have been satisfactorily settled, they are relatively easy to deal with. The impression that the main problems in cartography are centred on choice of symbols is quite incorrect. Although the exact balance of the different symbols in a complex map is vitally important, for the simple maps that are being considered here, well-established practices are relatively easy to describe.

The problem is that this kind of cartographic knowledge, driven by experience and imagination, is always specific. Given a different map, the questions and answers would be different. It does not proceed from general rules, because cartographers never deal with a theoretical 'map', only a specific one, and therefore they do not need to generalize any set of rules.

Although there have been a number of publications about the possible characteristics of cartographic expert systems for general use (Robinson and Jackson, 1985; Buttenfield and Mark, 1991), it has been realized that attempts to provide comprehensive rules for all the design decisions which may have to be taken in producing a complex multicolour map will require very complex systems indeed. Consequently, it is clear that most of the experimental expert systems which so far have been attempted in practice have tended to concentrate on thematic maps.

Broadly speaking, the problem of extracting general rules from cartographic experts is essentially avoided. The types of possible map to be produced are so limited that specific questions can be asked and answered. Although the examples described by Müller and Wang (1990) are clearly intended as demonstrations, it is evident that if the types of map to be produced from the database and the methods of representation and symbolization are sufficiently confined, there are virtually no 'design' questions left to answer. If the user simply has to command 'make a map of the variations in population density in the Netherlands by province,' and the database contains a suitable outline of the Netherlands and provincial boundaries, together with the required population data, the production could indeed be left to the program to perform. But if the user was to ask 'Would a map of the density of population in the Netherlands based on crude averages by province be of any real value?' the answer would have to be 'Only if you understand the limitations of mapping statistical data in this way.' Unfortunately, this is almost certainly what the inexpert user does not understand. Errors in design are visually evident and can be changed accordingly. Misjudgements in the choice of representational methods require far more attention in relation to statistical maps than they normally receive.

The fact that the requirements of the system may become unduly dominant is a problem recognized by developments in other fields. Bloomfield (1987), reviewing the use of Prolog in expert systems designed for legal purposes, points out that the use of Prolog 'has merely reflected the ways in which logic programmers see the world of law rather than the law as it actually is.'

In commenting on several attempts to produce cartographic expert systems, Forrest (1993: 145) makes this point clearly. As he says, 'This concentration on symbols rather than the presentation of phenomena must result in limitations on any general design system. Decisions about symbolization depend partly on what other information is included on the map, cannot be made in isolation, and cannot always be standardised.'

Representation and design

There is a general tendency to confuse the choice of a method of representation with the giving of form. The use of proportional symbols is a method: whether to use circles or squares, whether to show them as outlines or solid, and what colour to use, is a question of design. A contour is a method of representing elevation, but can also be used as an isoline to represent a specific value in any distribution or 'surface' which is continuously distributed (such as temperature). The gauge of the line and its hue are design decisions. Whereas the design decisions are essentially graphic, and can be approached quite differently in different maps using the same method of representation, the choice of representational method links the map to the phenomenon being represented.

It is the central issue of the representation of phenomena that lacks proper consideration. Without a much clearer understanding of this, so-called expert systems themselves are in danger of maintaining indifferent practice. When dealing with statistical maps in particular, the first requirement is to determine whether the phenomenon which the statistics purport to describe is continuous or discontinuous in its distribution. The proper choice of a method of representation must consider this in relation to the nature of the data available. Using choropleth maps to show population density averages is certainly well-established practice, but it is also bad practice. Such maps continue to be made, not because they are informative, but because they are simple to produce.

In many ways, the most effective expert system would concentrate on advising what sort of map, if any, can be made from the available data, and then leave the user with as few decisions as possible. Once the general plan of the map in terms of scale, area and content has been described, then the system will need to impose constraints on the unwary user. Given that such a user cannot anticipate the consequences of a decision, the visual demonstration of possible choices is very desirable. As many thematic maps use the same basic method of representation and encounter the same design issue for many different subjects, it is certainly possible to pre-program specific solutions. For example, if a scale of monochrome tints or a scale of colour tints with a certain number of steps is needed, then the actual specification of these can be built into the system, so that the user does not need to make design decisions which would be wrong or inadequate.

It is clear that investigations of expert systems of this type are only at the initial stage. Experience in other fields suggests that it is only when the problems involved are strictly limited that practical results can be achieved. As with other applications of artificial intelligence, the total replacement of the cartographer for all types of map is a long way off. It is also clear that if prototype systems are to be put into general use, it is essential that they are tested by inexperienced operators, and not just their authors.

Neural networks

The problems involved in trying to model human intelligence by using a serial digital computer have already been mentioned, and indeed some authors have maintained that as the human brain simply does not work by applying a set of rules, it will never be possible to make even a passable imitation of human thinking by any form of current computer programming. Advances in an alternative view are dependent on two related developments: a better understanding of how the brain works by forming neural networks; and the possibility of developing a different type of computer which is capable of 'learning' by forming equivalent networks.

In the earlier introductory description of visual perception, the importance of neural networks – the interconnections between neurons through synapses – has been stressed. The existence of such neural networks has been known for a long time, and attempts to measure or quantify them reveal a remarkable scale of complexity. Even in very approximate terms, there are about ten billion neurons in the cerebral cortex. The number of synaptic connections must be about one million billion. The variations in these connections reach truly astronomical proportions. In this sense alone the human brain is 'the most complicated material object in the known universe' (Edelman, 1994: 17).

Although the present emphasis is on the development of expert systems and other uses of artificial intelligence based on knowledge, there has been a resurgent interest in neural networks, and 'neural' computing, due largely to the remarkable advances in at least the theory of parallel computing. Aleksander (1989: 1) explains that this is based on the belief that 'the brain "computes" in a very different way from the conventional computer.' He maintains that the expert system approach to artificial intelligence is based on the premise 'that an understanding of what the brain does represents a true understanding only if it can be explicitly expressed as a set of rules that, in turn, can be run on a computer which subsequently performs artificially intelligent tasks.' He goes on to contrast this with the neural network approach, in which it is assumed that 'the brain, given sensors and a body, builds up its own hidden rules through what is usually called experience.' He continues, 'In neural computing it is believed that the cellular structures within which such rules can grow and be executed are the focus of important study as opposed to the AI concern of trying to extract the rules in order to run them on a computer.' Aleksander stresses that this does not lead to any attempt to build a machine which physically resembles the brain, but is concerned with the development of computer structures which are brain-like in operation.

The problem for the computer scientist is how to build a computer system which is capable of forming such connections on the basis of experience, just like a human being. The problems are immense. Whether it will be possible to construct such systems based on visual perception remains to be seen. According to Tattersall (1989: 41), 'a self-organizing array of neural elements with modifiable synaptic weights is capable of forming a topologically ordered map of data to which it has been exposed.' He explains that 'the neural map is generated by applying input patterns to the neural array without a human supervisor commanding the machine to assign the patterns to any particular class which the supervisor finds significant' (Tattersall, 1989: 43). Thus, theoretically, the machine 'learns' by itself, just as a human being does. Not surprisingly, this optimistic view of future possibilities is not shared by everyone. In reference to neural networks, or connectionist computers, Edelman (1994: 227) points out that ' "Neural nets" use symmetrical and dense matrixlike connections. In general they do not at all resemble the neuronal structures,' and 'neural networks are not adequate models or analogues of brain structure.'

Despite the remarkable possibilities being opened up by such research, there are still some interesting questions. The human brain is directly linked to its principal sense receptors, which feed information in, and can be directed by the brain to obtain information. Although the computer can be supplied with visual or aural information, it cannot be said to possess the equivalent of a visual system. It cannot perceive the external world. The system may learn to recognize a language, but could it respond by saying 'I don't like the tone of his voice'?

In addition, human beings live in a social and cultural world that both shapes their lives and is shaped by them. Activities such as making maps are intentional, and it is difficult to see how any machine can possess conscious intentions.

Man and machine

This whole research field has been taken a stage further in recent years by an increasing realization of the importance of neural networks, to the point where they are regarded as the most crucial aspect of the development of the human brain. This line of thought is now often referred to as neo-Darwinism or universal Darwinism, and it has great implications for a better understanding of both knowledge and skill. It is largely associated with the work of Edelman and associates.

An important element in these new ideas was established through research in immunology. According to Edelman (1994), the immune system can recognize foreign antibodies down to the molecular level, even if the molecules have been synthesized by organic chemists and have never existed before. It was this realization which destroyed the notion that the immune system was pre-programmed to deal with such events. Instead it was discovered that the individual has the ability to make a huge number of antibodies. In response to an invasion by a foreign molecule (such as on a virus), the antibodies with the best fit are then multiplied, and thus the population of cells is changed. This change is a selective adaptation to an intrusion from the external world, but it operates at the level of the human individual. Edelman regards this as paralleling the selective adaptation of a species taking place on the basis of evolutionary change of individual human beings, that is the genetic population.

In Edelman's words (1994: 78), the immune process has a recognition system which *'first* generates a diverse population of antibody molecules and then selects *ex post facto* those that fit or match.' This stresses the brain as a biological organ capable of performing selective recognition. It does not operate on the basis of previously existing instructions, and indeed could not because of the additional fact that the genetic code for each individual human does not result in producing uniform patterns at the level of neuronal circuits – they are unique for each individual in the species. Unlike a digital computer, there is no 'wiring diagram' to be followed. According to Edelman (1994: 25), 'the network of the brain is created by cellular movement during development, and by the extension and connection of increasing numbers of neurons. The brain is an example of a self-organizing system.' Each individual builds up circuits by selecting the most suitable or successful response from a large number of possible responses. Many of the basic operations are directed by hereditary requirements for the survival of the animal, but they still have to be developed within each individual organism.

It is impossible to do justice here to the full reach of these ideas, or the detailed scientific studies on which they are based. From a different point of view they are also systematically analysed by Plotkin (1994) in an eminently readable account of knowledge and intelligence. They help to explain why it is that the intelligent human can respond to new problems, not by bringing into action a set of rules, but by trying a number of different solutions or ideas until the most appropriate one is selected; and also why the development of conscious skills is achieved by beginning with a series of unsuccessful attempts which are gradually replaced by a practised ability. This ability is the result of reinforcing those neuronal circuits that have been proved to be most effective in reaching the goal, i.e. they are a selective adaptation. Therefore memory, an essential component not only of conscious learning and thought, but also of subconscious actions, is itself a consequence of forming neural networks which can be reactivated either by sensations or perceptions produced by external stimulus or by internal thoughts and feelings.

Because all these neural developments occur within individual human beings, any comprehensive theory has to explain why it is that different individuals have different memories, and that memories are also imperfect. Edelman (1994: 102) explains this by saying that 'recall is not stereotypic. Under the influence of continually changing contexts, it changes, as the structure and dynamics of the neural populations involved in the original categorization also change.' Because of the extensive interconnections, a given response may be achieved in several ways. In consequence, as Edelman (1994: 18) puts it, 'Unlike computer-based memory, brain-based memory is inexact, but it is also capable of great degrees of generalization.'

It is this diversity among human beings that makes possible their unequal capabilities. Although this is a problem where standardized and repetitive tasks are performed, it is an advantage in dealing with new sensations and new ideas. Even if a neural computer could be built to resemble the brain, it would of course be unique, and it would be impossible to construct another one just like it.

Given the remarkable complexity of the human brain, and the interaction of thought and feeling, of sensation and perception, it is essential that studies of the possible applications of artificial intelligence take this into account. At very least, it would suggest that at present it would be more rewarding to regard human beings and computers as being complementary rather than in opposition. It is also clear that the sorts of neural networks that are most relevant to both map production and map interpretation are only likely to be developed if they are formed and extended by practice in map-making and map-using tasks. Given the little attention they receive compared with language, it is only to be expected that many people rely most heavily on what can be articulated in language, which receives so much attention in both education and everyday life, and find difficulties with visual perception, which has not been so developed. The answer to this is indeed the one which many researchers have noticed in educating children in the use of maps. It is perfectly possible both to develop their interest and their comprehension, provided that sufficient time is devoted to it. This 'sufficient time' spent in both directed and exploratory attention is precisely what is needed to develop the neural networks on which perceptual skill and an understanding of maps can be built. Learning about maps is just as possible as learning about anything else. But it cannot he left to fragmentary and isolated opportunities. Such understanding requires the skilled use of visual perception, not just knowledge, and any improvement in both cartography and map interpretation has to be based on a recognition of this as a basic premise.

References

Akhmanova O 1977 *Optimization of natural communication systems.* Mouton, The Hague, Paris

Aleksander I 1989 Why neural computing? A personal view. In Aleksander I (ed) *Neural computing architectures.* North Oxford Academic, Oxford

Alison A 1811 *Essays on the nature and principles of taste.* Edinburgh

Anderson D R 1987 *Creativity and the philosophy of C S Peirce.* Martinus Nijhoff, The Hague

Anderson J R 1990 *Cognitive psychology and its implications*, 3rd edn. W H Freeman, Oxford

Andre G G 1891 *The draughtsman's handbook of plan and map drawing.* R and F Spon, London

Andrews J H 1975 *A paper landscape.* Oxford University Press, London

Aronoff S 1989 *Geographic information systems: a management perspective.* WDL Publications, Ottawa

Atlas Advanced 1968 Collins–Longman, Glasgow and London

Atlas Scolaire Suisse 1972 Editions Payot, La Conference des Chefs des Départements Cantonaux de l'Instruction Publique, Lausanne

Baddeley A D 1976 *The psychology of memory.* Harper and Row, New York

Bailey P 1984 The map in schools: a key, a language and a set of skills. *Cartographic Journal* **21**(1)

Baird J C 1970 *The psychophysical analysis of visual space.* Pergamon Press, Oxford

Beard M K 1991 Constraints on rule formation. In Buttenfield B P, McMaster R B (eds) *Map generalization.* Longman, Harlow

Bellin H 1971 The development of physical concepts. In Mischel T (ed) *Cognitive development and epistemology.* Academic Press, New York, London

Berleant A 1970 *The aesthetic field.* C C Thomas, Springfield

Bertin J 1978 Theory of communication and theory of the graphic. *International Yearbook of Cartography* **XVIII**

Black M 1975 *Caveats and critiques.* Cornell University Press, Ithaca

Blakemore C 1977 *Mechanics of the mind.* Cambridge University Press, Cambridge

Blocker H G 1979 *Philosophy and art.* Charles Scribner's Sons, New York

Bloomfield B 1987 Logic and rules of law. *The Guardian*, 26 March

Board C 1967 Maps as models. In Chorley R B, Haggett P (eds) *Models in Geography.* Methuen, London

Board C 1977 The geographer's contribution to evaluating maps as vehicles for communicating information. *International Yearbook of Cartography* **XVII**

Board C 1978a Map reading tasks appropriate in experimental studies in cartographic communication. *Canadian Cartographer* **15**(1)

Board C 1978b How can theories of cartographic communication be used to make maps more effective? *International Yearbook of Cartography* **XVIII**

Board C 1981 Cartographic communication. In Guelke L (ed) *Maps in modern geography. Cartographica* **18**(2) Monograph 27

Board C 1983 The development of concepts of cartographic communication with special reference to Professor Ratajsky. *International Yearbook of Cartography* **XXIII**

Boas F 1966 *Kwakiutl ethnography* (ed. Codere). University of Chicago Press, Chicago

Boden M 1987 *Artificial intelligence and natural man,* 2nd edn. The MIT Press, London

Bonin S 1975 *Initiation à la graphique.* Epi S. A. Editeurs, Paris

Brandes D 1976 The present state of perceptual research in cartography. *Cartographic Journal* **13**(2): 172–5

Brassel K E 1985 Strategies and data models for computer-aided generalization. *International Yearbook of Cartography* **XXV**

Brassel K E, Weibel R 1988 A review and conceptual framework of automated map generalization. *International Journal of Geographical Information Systems* **2**(3)

Brown R 1958 *Words and things.* The Free Press, New York

Bruce R L 1977 *Fundamentals of physiological psychology.* Holt, Rinehart and Winston, New York, London

Buttenfield B P 1991 A rule for describing line feature geometry. In Buttenfield B P, McMaster R B (eds) *Map generalization.* Longman, Harlow

Buttenfield B P, Mark D M 1991 Expert systems in cartographic design. In Taylor D R F (ed) *Geographic information systems: the microcomputer and modern cartography.* Pergamon Press, Oxford

Buttimer A 1983 *Creativity in context.* Lund Studies in Geography, series B 50, Royal University of Lund, C W K Gleerup

Capella J N 1972 The functional pre-requisites of intentional communicative systems. *Philosophy and Rhetoric* **5**(4)

Carline R 1968 *Draw they must.* Edward Arnold, London

Carswell R J B 1971 Children's abilities in topographic map reading. *Cartographica* **2**

Casey E S 1971–72 Expressionism and communication in art. *Journal of Aesthetics* **30**

Castner H A 1979a Viewing time and experience as factors in map design research. *Canadian Cartographer* **16**(2)

Castner H A 1979b A model of cartographic communication: practical goal or mental attitude? *International Yearbook of Cartography* **XIX**

Castner H A, Lywood D W 1978 Eye movement recording: some approaches to the study of map perception. *Canadian Cartographer* **15**(2)

Chang K-T, Antes J, Lenzen T 1985 The effect of experience and reading topographic map information: analyses of performance and eye movements. *Cartographic Journal* **22**(2)

Collier J 1890 *A manual of oil painting*. Cassell & Co, London

Coltheart M 1972 Memory. Introduction to Chapter Six. In Coltheart M (ed) *Readings in cognitive psychology*. Holt, Rinehart and Winston, Toronto

Coulson M R C 1977 Political truth and the graphic image. *Canadian Cartographer* **14**(2)

Craig W M 1821 *A course of lectures on drawing, painting and engraving*. Longman Hurst, London

Crick M 1976 *Explorations in language and meaning*. Malaby Press, London

Crowther P 1982 Merleau-Ponty: perception into art. *British Journal of Aesthetics* **22**(2)

Cuenin R 1972 *Cartographie Générale*. 2 vols. Eyrolles, Paris

Culler J 1976 *Saussure*. Fontana, London

Dainville F de 1964 Langage des Géographes. Editions A & J Picard, Paris

De Grote Bosatlas 1971 Walters–Noordhoff, Groningen

Ditchburn R W 1973 *Eye movements and visual perception*. Clarendon Press, Oxford

Dixon P 1971 *Rhetoric*. Methuen, London

Dobson M W 1977 Eye movement parameters and map reading. *American Cartographer* **4**(1)

Dobson M W 1979 Visual information processing during cartographic communication. *Cartographic Journal* **16**(1): 79

Dowling J E, Boycott B B 1966 Organisation of the primate retina. *Proceedings Royal Society London Series B* **166**

Dreyfus H L 1986 Why studies of human capacities modelled on ideal natural science can never achieve their goal. In Margolis J, Krausz M, Burian R M (eds) *Rationality, relativism and the human sciences.* Martinus Nijhoff, The Hague

Dreyfus H L, Dreyfus S E 1988 Making a mind versus modelling the brain. In Graubard (ed) *The artificial intelligence debate.* MIT Press, Cambridge

Ducasse C J 1968 *Truth, knowledge and causation.* Routledge and Kegan Paul, London, New York

Eastman J R 1985 Cognitive models and cartographic design research. *Cartographic Journal* **22**(2)

Eco U 1976 *A theory of semiotics.* Indiana University Press, Bloomington

Eco U 1985 Producing signs. In Blonsky M (ed) *On signs,* Basil Blackwell, Oxford

Edelman G 1994 *Bright air, brilliant fire.* The Penguin Press, London,

Eldridge R 1985 Form and content: an aesthetic theory of art. *British Journal of Aesthetics* **25**(4)

Ellis H C 1972 *Fundamentals of human learning and cognition.* W C Brown, Dubuque

Fellows B J 1968 *The discrimination process and development.* Pergamon Press, Oxford

Fisher A B 1978 *Perspectives on human communication.* MacMillan, New York, London

Fisher P, Dykes J, Wood J 1993 Map design and visualisation. *Cartographic Journal* **30**(2)

Forgus R H, Melamed L E 1976 *Perception: a cognitive stage approach.* McGraw–Hill, New York

Forrest D 1993 Expert systems and map design. *Cartographic Journal* **30**(2)

Gaarder K R 1975 *Eye movements, vision and behaviour.* Wiley, New York

Geach P 1957 *Mental acts.* Routledge and Kegan Paul, London

George F H 1962 *Cognition.* Methuen, London

Gilbert J 1885 *Landscape in art*. John Murray, London

Gill G 1993 Road map design and route selection. *Cartographic Journal* **30**(2)

Gilmartin P 1992 Twenty-five years of cartographic research: a content analysis. *Cartography and Geographic Information Systems* **19**(1)

Gogel W C 1968 The measurement of perceived size and distance. In Neff W D (ed) *Contributions to sensory psychology*, Vol. 3. Academic Press, New York, London

Gombrich E 1977 *Art and illusion*, 5th edn. Phaidon Press, London

Goodman N 1969 *Languages of art*. Oxford University Press, London

Gould J D, Dill A B 1969 Eye movement parameters and pattern discrimination. *Perception and Psychophysics* **6**(5)

Gray B 1977 *The grammatical foundations of rhetoric*. Mouton, The Hague

Green D 1993 Map output from geographic information and digital image processing systems. *Cartographic Journal* **30**(2)

Greenlee D 1973 *Peirce's concept of sign*. Mouton, The Hague

Gregory R L 1974 *Concepts and mechanisms of perception*. Duckworth, London

Guelke L 1977 Cartographic communication and geographic understanding. *Cartographica*, Monograph 19

Hagen M A 1986 *Varieties of realism*. Cambridge University Press, Cambridge

Hamerton P G 1882 *The graphic arts*. Seeley, Jackson and Halliday, London

Hamlyn D W 1957 *The psychology of perception*. Routledge and Kegan Paul, London, New York

Harley J B 1989 Deconstructing the map. *Cartographica* **26**(2)

Harley J B 1990 Cartography, ethics and social theory. *Cartographica*, **27**(2)

Harley J B 1991 Can there be a cartographic ethics? *Cartographic Perspectives* **10**

Hartley A A 1977 Mental measurement in the estimation of length. *Journal of Experimental Psychology (Human Perception and Performance)* **3**(4)

Harvey D 1969 *Explanation in geography.* Edward Arnold, London

Hausman C R 1981 Criteria of creativity. In Dutton D, Krausz M (eds) *The concept of creativity in science and art.* Martinus Nijhoff, The Hague

Head C G 1984 The map as natural language: a paradign for understanding. *Cartographica,* Monograph 31

Hearnshaw H M, Unwin D J 1994 (eds) *Vizualisation in geographical information systems.* John Wiley & Sons, Chichester

Heelan P A 1983 *Space perception and the philosophy of science.* University of California Press, Los Angeles

Hendrikson H 1975 The map as an idea: the role of cartographic imagery during the Second World War. *American Cartographer* **2**(1)

Herriot P 1974 *Attributes of memory.* Methuen, London

Hill A R 1974 *Cartographic performance: an evaluation of orthophotomaps.* US Army Engineer & Topographic Research Labs, Washington DC

Hill A R and Burns J R 1978 *Towards a measure of cartographic competence.* 9th International Cartographic Conference, International Cartographic Association, Washington DC

Hocking D, Keller C P 1992 A user perspective on atlas content and design. *Cartographic Journal* **29**(2)

Hookway C 1985 *Peirce.* Routledge and Kegan Paul, London, New York

Horridge G A 1968 *Interneurons.* W H Freeman, San Francisco

Hospers J 1985 Artistic creativity. *British Journal of Aesthetics and Art Criticism,* Spring

Howard H 1848 *A course of lectures on painting.* Henry G Bohn, London

Howell W S 1971 *Eighteenth-century British logic and rhetoric.* Princeton University Press, Princeton

Hubel, D H, Wiesel, T N 1968 Receptive fields and functional architecture of the monkey striate cortex. *Journal of Physiology* **195**

Hübner K 1983 Dixon P R jnr, Dixon H M (transl.) *Critique of scientific reason.* University of Chicago Press, Chicago

Hurvich L M, Jameson D 1957 An opponent-process theory of colour vision. *Psychological Review* **64**

Hurvich L M, Jameson D 1974 Opponent processes as a model of neuronal organisation. *American Psychologist* **29**

Imhof E 1951 *Terrain et carte*. Eugen Rentsch Verlag, Zürich

Imhof E 1962 Die anordung der namen in der karte. *Internationales Jahrbuch für Kartgraphie* **II**: 93–126

Imhof E 1965 *Kartographische geländedarstellung*. Walter de Gruyter, Berlin. Also 1982 Steward J J (transl. ed.) *Cartographic relief presentation*. Walter de Gruyter, Berlin

Innis R E 1985 *Semiotics: an introductory anthology*. Indiana University Press, Bloomington

James T G H 1985 *Egyptian painting and drawing*. British Museum, London

Jannace R, Ogrosky C 1987 Cartographic programs and products of the United States Geological Survey. *American Cartographer* **14**(3)

Johnston D M 1972 *Systematic introduction to the psychology of thinking*. Harper and Row, New York

Johnstone H W 1975 *Validity and rhetoric in philosophical argument*. Dialogue Press, New York

Kain R J P 1975 R K Dawson's proposal in 1836 for a cadastral survey of England and Wales. *Cartographic Journal* **12**(2)

Kanakubo T, Morita T 1993 Introduction: the selected main theoretical issues facing cartography (ICA Report). *Cartographica* **30**(4)

Karssen A J 1980 Design research (letter). American Cartographer 7(1)

Katz I 1994 Checkmate for the superchip. *The Guardian*, 2 September

Keates J S 1989 *Cartographic design and production*, 2nd edn. Longman, Harlow and John Wiley & Sons, New York

Kennedy G 1969 *Quintilian*. Twayne Publications, New York

Keuss P J G 1977 Processing of geometrical dimensions in a binary classification task. *Psychology and Perception* **21**(4)

Kittay E F 1987 *Metaphor*. Clarendon Press, Oxford

Klausmeier H J, Ghatala E S, Frayer D A 1974 *Conceptual learning and development*. Academic Press, New York

Koláčný A 1969 Cartographic information: a fundamental concept and term in modern cartography. *Cartographic Journal* **6**(1)

Koláčný A 1971 The importance of cartographic information for the comprehending of messages spread by mass communication media. *International Yearbook of Cartography* **XI**

Kolers P A 1973 Some modes of representation. In Pliner P, Krames L, Alloway T (eds) *Communication and affect*. Academic Press, New York, London

Laidlay W J 1911 *Art, artists and landscape painting*. Longman Green, London

Langer S K 1953 *Feeling and form*. Scribners

Langer S K 1957 *Problems of art*. Routledge and Kegan Paul, New York

Langran G 1991 Generalization and parallel computation. In Buttenfield B P, McMaster R B (eds) *Map Generalization*. Longman, Harlow

Lanigan R L 1972 *Speaking and semiology*. Mouton, The Hague

Lee D 1992 *Cartographic Generalization*. Mapping Sciences Division, Intergraph Corporation, Huntsville

Leibovic K N 1972 *Nervous system theory*. Academic Press, London, New York

Loxton J 1985 The Peters phenomenon. *Cartographic Journal* **22**(2)

Lyne J R 1982 C S Peirce's philosophy of rhetoric. In Vickers B (ed.) *Rhetoric revalued*. International Society for the History of Rhetoric Monograph 7, State University of New York, Binghampton

MacCormac E R 1971–72 Metaphor revisited. *Journal of Aesthetics and Art Criticism* **30**

MacEachren A M 1991 The role of maps in spatial knowledge acquisition. *Cartographic Journal* **28**(2)

McGrath G 1965 The representation of vegetation on topographic maps. *Cartographic Journal* **2**(2)

McGrath G 1971 The mapping of national parks: a methodological approach. *Cartographica* **2**

McMaster R B 1991 Conceptual frameworks for geographic knowledge. In Buttenfield B P, McMaster R B (eds) *Map generalization*. Longman, Harlow

McNally J R 1970 Toward a definition of rhetoric. *Philosophy and rhetoric* **3**(2)

McQuail D 1975 *Communication*. Longman, London, New York

Mackaness W A 1991 Integration and evaluation of map generalization. In Buttenfield B P, McMaster R B (eds) *Map generalization*. Longman, Harlow

Maling D H 1973 *Coordinate systems and map projections*. George Philip, London

Meeson P 1972 Drawing, art and education. *British Journal of Aesthetics* **12**

Mendelowitz D M 1970 *A history of American art*. Holt, Rinehart and Winston, New York

Michie D 1989 A new neural fever to use our noggins. *The Guardian*, 16 March

Miller L L 1973 *Knowing, doing and surviving*. Wiley, New York, London

Miller R W 1987 *Fact and method*. Princeton University Press

Moles A 1968 *Information theory and esthetic perception*. Cohen J E (transl) University of Illinois Press, Urbana, Chicago

Morris C 1971 *Writings on the general theory of signs*. Mouton, The Hague

Morris J 1982 The magic of maps: the art of cartography. MA thesis, University of Hawaii

Morrison A 1979 The frequency of map use for planning journeys by road. *Society of University Cartographers Bulletin* **13**(1)

Morrison A 1980 Existing and improved road classifications for British road maps related to the speed of traffic. *Cartographic Journal* **17**(2)

Morrison J L 1974 A theoretical framework for cartographic generalisation with emphasis on the process of symbolisation. *International Yearbook of Cartography* **XIV**

Morrison J L 1976 The science of cartography and its essential processes. *International Yearbook of Cartography* **XVI**

Morrison J L 1978 Towards a functional definition of the science of cartography with emphasis on mapreading. *American Cartographer* **5**(2)

Muehrcke P C 1972 *Research in thematic cartography. Resource Paper 19.* Commission on College Geography, Association of American Geographers, Washington DC

Mukařovský J 1978 *Structure, sign and function. Selected essays.* Burbank J, Steiner P (transl). Yale University Press, New Haven

Müller J C, Wang Zeshan 1990 A knowledge-based system for cartographic symbol design. *Cartographic Journal* **27**(1)

Munro C F 1987 Semiotics, aesthetics and architecture. *British Journal of Aesthetics* **27**(2)

Munro T 1946 Style in the arts: a method of stylistic analysis. *Journal of Aesthetics and Art Criticism* **5**(2)

Murray E L 1986 *Imaginative thinking and human existence.* Duquesne University Press, Pittsburgh

Navon D 1977 Forest before trees: the presence of global features in visual perception. *Cognitive Psychology* **9**

Neisser U 1976 *Cognition and reality.* W H Freeman, San Francisco

Nickerson B G 1991 Knowledge engineering for generalization. In Buttenfield B P, McMaster R B (eds). *Map generalization.* Longman, Harlow

Norman D A, Rumelhart D E 1975 *Explorations in cognition.* W H Freeman, San Francisco

Nørreklit L 1973 *Concepts: their nature and significance for metaphysics and epistemology.* Odense University Press, Odense

Oatley K 1978 *Perceptions and representations.* Methuen, London

Olson J M 1975 Experience and the improvement of cartographic communication. *Cartographic Journal* **12**(2) 94–108

Orr R A 1991 Bartholomew and the giant GIS. *Cartographic Journal* **28**(1)

Osaka N 1980 Effect of peripheral visual field size upon visual search in children and adults. *Perception* **9**(4)

Osborn J 1743 *The principles of painting by a painter.* London

Osborne H 1986 Interpretation in science and in art. *British Journal of Aesthetics* **20**(1)

Paivio A 1971 *Imagery and verbal processes*. Holt, Rinehart and Winston, New York

Pannekoek A J 1962 Generalisation of coastlines and contours. *International Yearbook of Cartography* **II**

Parkhurst D B 1898 *The painter in oil*. Boston

Paul D H 1975 *The physiology of nerve cells*. Blackwell, Oxford

Peirce C S 1934, 35, 65 *Collected Papers*. Hartshorne C, Weiss P (eds). Harvard University Press

Perkins D, Leondar B 1977 Symbol systems: the stuff of art. In Perkins, Leondar (eds). *The Arts and Cognition*. Johns Hopkins University Press, Baltimore

Petchenik B B 1974 A verbal approach to characterising the look of maps. *American Cartographer* **1**(1)

Petchenik B B 1977 Cognition in cartography. *Cartographica* **19**

Petchenik B B 1979 From place to space: the psychological achievement of thematic mapping. *American Cartographer* **6**(1)

Petchenik B B 1983 A map-maker's perspective on map design research 1950–80. In Taylor D R F (ed) *Progress in Contemporary Cartography*, vol. 2. Wiley, New York

Petchenik B B 1985 Maps, markets and money. *Cartographica* **22**(3)

Piaget J. 1971 *Biology and knowledge*. Edinburgh University Press, Edinburgh

Pickles J 1985 *Phenomenology, science and geography*. Cambridge University Press, Cambridge

Piket J J C 1972 Five European topographic maps. *Geografisch Tijdschrift* **6**(3)

Pirenne M H 1967 *Vision and the eye*. Chapman and Hall, London

Pitkänen R 1980 On the analysis of pictorial space. *Acta Philosophica Fennica* **31**(4)

Plotkin H 1994 *The nature of knowledge*. The Penguin Press, London

Polyani M 1969 *Knowing and being*. Routledge and Kegan Paul, London

Pope A 1949 *The language of drawing and painting*. Harvard University Press, Cambridge

Price H H 1981 Painting and the theory of knowledge. *British Journal of Aesthetics* **21**

Pye D 1978 *The nature of aesthetics and design*. The Herbert Press, London

Pylyshyn Z W 1973 What the mind's eye tells the brain: a critique of mental imagery. *Psychological Bulletin* **80**

Ranzinger M 1985 A data structure for a geo-expert system knowledge base. *International Yearbook of Cartography* **XXV**

Ratajsky L 1973 The research structure of theoretical cartography. *International Yearbook of Cartography* **XIII**, also *Cartographica* **19** (1977)

Reeke G N, Edelman G M 1988 Real brains and artifical intelligence. In Graubard S R (ed) *The artificial intelligence debate*. MIT Press, Cambridge

Reeves E A 1910 *Maps and map making*. Royal Geographical Society, London

Reid L A 1969 *Meaning in the arts*. George Allen and Unwin, London

Rescher N 1973 *Conceptual idealism*. Blackwell, Oxford

Richards I A 1936 *The philosophy of rhetoric*. Oxford University Press, London

Rieger M K, Coulson M R C 1993 Consensus or confusion: cartographers' knowledge of generalization. *Cartographica* **30**(2/3)

Roberts R 1984 Aristotle: the rhetoric. In Barnes J (ed) *The complete works of Aristotle*, vol. 2. Princeton University Press

Robinson A H, Petchenik B B 1975 The map as a communication system. *Cartographic Journal* **12**(1)

Robinson A H, Petchenik B B 1976 *The nature of maps*. University of Chicago Press, Chicago, London

Robinson A H, Sale R D, Morrison J L 1978 *Elements of cartography*, 4th edn. Wiley, New York

Robinson G, Jackson M 1985 Expert systems in map design. *Proceedings, Autocarto* **7**

Rose S 1986 Introduction: In Rose S, Appignanesi L (eds) *Science and beyond*. Basil Blackwell, Oxford

Rosenfield L W 1971 *Aristotle and information theory*. Mouton, The Hague

Royal Institution of Chartered Surveyors 1980 *Specifications for mapping at scales between 1:1000 and 1:10 000*. RICS, London

Russell B 1914 *Our knowledge of the external world*. The Open Court Publishing Company, Chicago and London; and 1972 G Allen and Unwin, London

Rystedt B 1977 The Swedish land data bank – a multipurpose information system. *Cartographica*, Monograph 20

Salichtchev K A 1973 Some reflections on the subject and method of cartography after the Sixth International Cartographic Conference. *Canadian Cartographer* **10**(2)

Salichtchev K A 1978 Cartographic communication: its place in the theory of science. *Canadian Cartographer* **15**(2)

Salkie R 1992 *The Chomsky Update*. Routledge, London, New York

Salomon L B 1966 *Semantics and common sense*. Holt, Rinehart and Winston, New York, London

Sandford H A 1980 Directed and free search of the school atlas map. *Cartographic Journal* **17**(2)

Sandford H A 1985 The future of the school pupils' desk atlas. *Cartographic Journal* **22**(1)

Schaper E 1964 The art symbol. *British Journal of Aesthetics* **4**

Schaper E 1972 *Style names and the concept of style*. Proceedings 6th International Congress on Aesthetics (Uppsala 1968), Acta Universitatis Upsaliensis, figura nova Series 10

Schwarz J T 1988 The new connectionism. In Graubard S R (ed) *The Artificial Intelligence Debate*. MIT Press, Cambridge

Schlichtmann H 1985 Characteristic traits of the semiotic system 'map symbolism'. *Cartographic Journal* **22**(1)

Scruton R 1974 *Art and imagination*. Methuen, London

Sebeok T A 1976 *Contributions to the doctrine of signs.* Indiana University Press, Bloomington

Shepard R N, Metzler J 1971 Mental rotation of three-dimensional objects. *Science* **171**

Skelton R A 1962 The origins of the Ordnance Survey of Great Britain. *Geographical Journal* **128**

Sless D 1986 *In search of semiotics.* Croom Helm, London

Solman R T 1975 Relationship between selection accuracy and exposure in visual search. *Perception* **4**

Sparshott F E 1981 Every horse has a mouth: a personal poetics. In Dutton D, Krausz M (eds) *The concept of creativity.* Martinus Nijhoff, The Hague

Sperling G 1960 The information available in brief visual presentation. *Psychological Monographs* **74**(11)

Sprigge T L S 1970 *Facts, words and beliefs.* Routledge and Kegan Paul, New York

Steer F W 1967 A dictionary of land surveyors in Britain. *Cartographic Journal* **4**(2)

Steiner W 1982 *The colors of rhetoric.* University of Chicago Press, Chicago

Stevens S S 1975 *Psychophysics: introduction to its perceptual, neural and social aspects.* Stevens G (ed), Wiley, New York, London

Sutherland N S 1973 Object recognition. In Carterette E C, Friedman M P (eds) *Handbook of perception III*, Academic Press, New York, London

Tattersall G 1989 Neural map applications. In Aleksander I (ed) *Neural computing architectures.* North Oxford Academic, Oxford

Taylor D R F (ed) 1983 *Introduction: graphic communication and design in contemporary cartography.* John Wiley & Sons, New York

Taylor D R F 1993 Geography, GIS, and the modern mapping sciences: convergence or divergence. *Cartographica* **30**(2/3)

Taylor R M 1974 An evaluation of an experimental map for the projected map display. 7th International Cartographic Conference, International Cartographic Association, Madrid

Teghtsoonian M 1965 The judgment of size. *American Journal of Psychology* **78**

Thorndyke P W, Stasz C 1980 Individual differences in procedures for knowledge acquisition from maps. *Cognitive Psychology* **12**

Torok Z 1993 Social context. In *The selected main theoretical issues facing cartography*, ICA Report, *Cartographica* **30**(4)

Tyler S A 1986 Post-modern anthropology. In Chock P P, Wyman J R (eds) *Discourse and the social life of meaning*. Smithsonian Institution Press, Washington

Vibert J G 1892 *The science of painting*. Percy Young, London

Vickers D 1979 *Decision processes in visual perception*. Academic Press, New York

Vickers D 1982 Territorial disputes: philosophy v rhetoric. In Vickers D (ed) *Rhetoric revalued*. Monograph 1, International Society for the History of Rhetoric, State University of New York, Binghampton

Visvalingam M, Whyatt J D 1993 Line generalisation by repeated elimination of points. *Cartographic Journal* **30**(1): 46–51

Waltz D L 1988 The prospects for building truly intelligent machines. In Graubard S R (ed) *The artificial intelligence debate*. MIT Press, Cambridge

Washburn B 1960 *Mt McKinley, Alaska, 1 : 50 000*. Museum of Science, Boston

Washburn B 1978 *The Heart of the Grand Canyon, 1 : 24 000*. National Geographic Society in collaboration with the Museum of Science, Boston

Wastesson O 1977 Computer cartography and GIS in Sweden: a synopsis. *Cartographica*, Monograph **20**

Watson J 1982 Tectonics: moving with the times. *Nature* **300**

Weibel R 1991 Amplified intelligence and rule-based systems. In Buttenfield B P, McMaster R B (eds) *Map generalization*. Longman, Harlow

White J 1957 *The birth and rebirth of pictorial space*. Faber and Faber, London

Whittlesey D (undated) *German strategy of world conquest.* F E Robinson and Co

Wollheim R 1987 *Painting as an art.* Thames and Hudson, London

Wolterstorff N 1980 *Works and worlds of art.* Clarendon Press, Oxford

Wood C H 1985 Tonal reproduction processes in map printing from the 15th to the 19th centuries. *Cartographica* **22**(1)

Wood D, Fels J 1986 Designs on signs: myth and meaning in maps. *Cartographica* **23**(3)

Wood M 1993 The user's response to map design. *Cartographic Journal* **30**(2)

Woodfield R 1986 Words and pictures. *British Journal of Aesthetics* **26**(4)

Wright I D M 1978 Thematic maps: aspects of their visual effectiveness. Unpublished thesis, University of Dundee, Dundee

Wyszecki G, Stiles W S 1982 *Color science.* J. Wiley & Sons, New York

Yonge C D 1852 *The Orations of M T Cicero,* vol. 7. London

Young J Z 1978 *Programs of the brain.* Oxford University Press, Oxford

Index